T0146323

Learning from Construction Failures: Applied Forensic Engineering

Learning from Construction Failures: Applied Forensic Engineering

edited by

Peter Campbell

Consultant and Founding Partner, Campbell Reith Hill

Whittles Publishing

Typeset by
Whittles Publishing Services

Published by
Whittles Publishing,
Roseleigh House,
Latheronwheel,
Caithness, KW5 6DW,
Scotland, UK

© 2001 Peter Campbell

All rights reserved.
No part of this publication may be reproduced,
stored in a retrieval system, or transmitted,
in any form or by any means, electronic,
mechanical, recording or otherwise
without prior permission of the publishers.

ISBN 1-870325-63-X

The publisher assumes no responsibility for any injury and/or damage to persons or property from the use or implementation of any methods, instructions, ideas or materials contained within this book. All operations should be undertaken in accordance with existing legislation and recognised trade practice. Whilst the information and advice in this book is believed to be true and accurate at the time of going to press, the authors and publisher accept no legal responsibility or liability for errors or omissions that may be made.

Printed by Bookcraft Ltd., Midsomer Norton, UK

Contents

Examples and case studies

Foreword

Engineers perhaps have the reputation of getting on with the next job without looking back. Indeed, in the present day, the pressure exerted by the fee structure militates, so to speak, against extraneous work. There is no place or fee for a desk study except in rare cases. The present day easy access through the Internet to large amounts of data and to case studies will, I suspect, make little difference to this situation. There seems to be an inherent need to reinvent wheels; the cycle for this appears to shorten as the decades pass. In that excellent book *The Wheelwright's Shop,* George Sturt says that every successful wheelwright 'must chop his knee'. We know what he means. Yet we also know, deep down, that this is nonsense and indeed buildings get taller and bridges get longer, or is this learning from our successes? Nevertheless it is often the 'simpler' things that go wrong; the experience of using materials and processes does not permeate sufficiently widely nor is it sufficiently understood and this can lead to failures in the eyes of the client.

Failure is an emotive word and means many different things, but I believe in all cases it means that the client has had an unsatisfactory solution to his problem. In some cases the failures are catastrophic, for example the Tacoma Narrows Bridge, but in many it might simply be unsightly cracking.

One of the reasons for this state of affairs is perhaps, curiously enough, the code of practice mentality which inhibits the holistic approach to 'design' and which has come more and more to mean analysis. In my visits to university schools of engineering I have been perturbed by the introduction of codes into courses although I am aware of the reasons for it. This view is not to decry codes which without doubt have been the means of avoiding failures in many cases. In codes the whole is broken down into parts which are analysed separately. The result is a safe structure but not one in which the strains and stresses have much semblance to the calculated ones.[1] Again buildings are designed for self and applied loads but rarely for the strains caused for example by shrinkage and creep of concrete. National Building Studies Research Paper No 28 records that the biggest change of strain in the steel beams of the Ministry of Defence Building in Whitehall was caused by the shrinkage of the concrete after the floors had been cast – this was larger than that induced by the loading of the floors subsequently. At present we know so little about the behaviour of buildings as opposed to that of their constituent parts although some progress has been made in recent years in the trials at the BRE Large Building Test Facility at Cardington.

The other great problem is the legal one, intertwined as it is with insurance. Until a way is found through this morass, failures as such are likely to be but dimly seen. Nevertheless it is sensible and desirable from time to time to collect such evidence as there is of 'failures' and the reasons for them as far as can be ascertained and to publish them for the erudition of practitioners in this field and particularly those who are beginning their career. Only rarely will we learn from our successes. Even if the failures are ones which we fondly think we would not cause, nevertheless there are likely to be similarities which should cause us to re-examine the work which we are doing. Some 'famous' failures such as the steel box girder bridges are not covered in this book – they have been adequately covered elsewhere[2] – but they do warn us when we 'copy' a design in one material but using another one, that we must ensure that the inherent and perhaps incalculable properties of the one are present in the other.

Or again we must ensure as far as possible that the manufacturing or construction process itself, which is not often taken into account in 'design', does not set up situations which can be disastrous. It has been known for a welder striking his arc on a piece of steelwork to set up a notch brittle situation which gave rise to complete failure of a structure under low temperature. For one who has had to design and test structures for extreme conditions such as high explosive and nuclear attack it has been salutary to learn that a structure will do its utmost not to fall down. It will find paths for the load to follow which the engineer might find inconceivable. However, correspondingly local failures which could have been avoided with different detailing have contributed adversely to the overall behaviour of structures.

What I hope readers, particularly those who have just entered the profession, will find in this collection is not only clues to avoiding failure but also the need to stand back on completion of an analysis and to try and visualise how material properties might change with time; how loads will be carried safely to the ground not only through the idealised structure they have used in their analysis but also through other additions which might help the overall stability or alternatively prevent the structure behaving in the way anticipated. I suppose some would say that this approach is akin to risk analysis. Risk presupposes failure. I consider it is facing reality. The only definition of a professional which I am prepared to entertain is someone who uses his/her judgement and not a set of rules. This book with its wide range of authors and contributions should help to mature that judgement.

Dr Francis Walley, CB, FREng

---

[1] Conference on the correlation between calculated and observed stresses and displacements in structures, ICE 1956.
[2] Criteria for the assessment of steel box girders with particular reference to the bridges at Milford Haven and Avonmouth 1970, Department of the Environment.

Preface

Generally in the developed world, the design and the construction of all aspects of the built environment are regulated by rules and procedures that are designed to minimise the risk of failures, in the interests of public safety and for all those that use those facilities.

It is therefore not surprising that the incidence of significant failures, in relation to the design and construction of the existing estate and the volume of construction that is procured each year, is thankfully small. However, this reality is not reflected on a global basis, even in parts of the developed world.

Then there are the spectacular failures that are caused by natural disasters such as earthquakes, cyclones and floods, which are difficult to account for when designs are carried out, particularly if the promoter requires the expenditure to be kept to a minimum. It is nevertheless possible to generate a high level of protection in such circumstances, provided the design concept and its technical implementation reflects the relevant technology.

Failure is an emotive word, and in the litigious world in which we live, it is understandable that when these problems arise, those responsible want to keep the details as private as possible. This is, however, contrary to the best interests of those that design, specify and construct buildings and structures, because it is in these circumstances, when things have gone wrong and have to be carefully investigated, that we can learn the important lessons that will limit the recurrence of similar events in the future.

In 1867 the architect Edwin Nash presented a paper at the Royal Institution of British Architects entitled *Remarks Upon Failures In Construction*, and he said this:

> *Architects would do well to preserve the utmost amount of detail when failures happen, such particulars being most important aids in promoting a correct knowledge of constructive principles; and, although it is natural to be silent upon mishaps, it is not always necessary to be so ... There are occasions in which architects may beneficially exhibit their* esprit de corps *in a revelation of the difficulties and results of their practice, for it is often the mystery which envelops a subject, and not the full statement of the truth, that does harm to those concerned.*

That statement is as true today as it was over 130 years ago and applies to all those involved in the construction process.

Education is a fundamental aspect for the preparation of people who wish to become a part of the construction industry; and from my experience civil engineering education concentrates on design analysis and generally devotes little or no time to engineering concepts that are ultimately subjected to analysis. Statistically about 30% of all failures result from design errors, and I use the word design in its broadest sense.

A crucially important aspect of design is to consider at each stage what could possibly go wrong if certain things happened, right through to the completion of the construction. Such systematic reviews during the whole procurement process may well signpost appropriate modifications in the interests of the lifetime performance of the facility in question, but this approach is rarely to my knowledge examined in engineering courses.

Quality systems worthy of the name require that all aspects of the design are thoroughly and independently checked. Failure to follow this principle is to court disaster and the possibility of a ruined career, especially when loss of life occurs. Engineers, like doctors, have a responsibility to protect lives by ensuring through the quality of everything they do, that failures which may put lives at risk do not occur.

This book has been written by a number of distinguished experts, with the primary objective of it being used as an educational tool. It focuses on the widest aspects of civil and structural engineering design and construction, and Chapters 1, 2 and 3 set out the historical perspective including the process of forensic investigation as practised in North America.

Chapters 4, 5 and 6 describe the application of the principles as applied to failure investigations including a fascinating account of the work that has been carried out to stabilise the Tower of Pisa. Chapters 7 to 10 deal with the important issues of insurance and the legal aspects of failure investigations. These chapters include such subjects as risk assessment and management issues, with guidance from experienced colleagues both here and in the USA, on how best to limit the incidence of failures and with the use of case studies, how failures have been investigated and remedial works carried out.

The remaining chapters deal with a series of case studies which have a distinctly international flavour. They cover modern buildings and structures and those that can be truly described as historic, from reservoir control structures to the failure of masts and towers, and in graphic detail, the Oklahoma bombing atrocity.

The depth and breadth of the case studies are impressive, taking the reader to the very broadest reaches of construction – and each one has something to offer, something to say – if only we are willing to read and accept and learn from the experience of knowledgeable failure investigators.

I am indebted to all the authors, for the expression of their collective knowledge and experience, which I am confident will be of great value to those who read this book, both students and practitioners alike.

Peter Campbell

Contributors

Mr. Poul Beckmann FIStructE, MICE, MIDA, HonFRIBA
>65 Chartfield Avenue, London, SW15 6HN, formerly Consultant, Ove Arup & Partners, 13 Fitzroy Street, London, W1T 4BQ

Professor John Burland DSc (Eng), FRS, FREng,
>Professorof Soil Mechanics, Civil Engineering Department, Imperial College, London, SW7 2BU

Mr. J.S.Carlton CEng, MIMechE, MIMarE, MRINA
>Lloyd's Register, Technical Investigation Department, 71 Fenchurch Street, London, EC3M 4BS

Professor Kenneth L. Carper Architect, MASCE
>School of Architecture and Construction Management, Washington State University, P.O. Box 642220, Pullman, WA 99164-2220, USA

Dr. John Chapman FREng, FCGI, PhD, FIStructE, FICE, FRINA
>41 Oathall Road, Haywards Heath, Sussex, RH16 3EG and Visiting Professor, Civil and Environmental Engineering Department, Imperial College, London, SW7 2BU

Dr. W. Gene Corley PhD, MNAE, FASCE, FACI, FIStructE
>Senior Vice President, Construction Technology Laboratories, 5420 Old Orchard Road, Skokie, Illinois IL 60077, USA

Dr. Satish Desai, OBE, BE, PhD, FIStructE
>Visiting Professor, School of Engineering in the Environment, University of Surrey, Guildford, GU2 7XH

Ms. Diana Holtham LLB
>formerly Head of Construction Division, Berrymans Lace Mawer Solicitors diana.holtham@virgin.net

Mr. Lawrance Hurst BScEng, FCGI, CEng, FICE, FIStructE, FConsE, FBEng
  Consultant to Hurst, Peirce & Malcolm, Celtic House, 33 John's Mews, London, WC1N 2QL

Professor A. Kennaway CEng, FIMechE, FPRI
 (deceased)Engineering Consultant, 12 Fairholme Crescent, Ashtead, Surrey, KT12 2HN

Professor J. Lewin (Hon) DEng, CEng, FICE, FIMechE, FCIWEM
  2 Beechrow, Ham Common, Richmond, Surrey. TW10 5HE

Dr. John Maguire CEng, FICE, FIStructE, MIMarE
  Lloyd's Register, Technical Investigation Department, 71 Fenchurch Street, London, EC3M 4BS

Mr. Robert A. Rubin PE, FASCE admitted to bar in New York
  Postner & Rubin Attorneys at Law, 17 Battery Place, New York, New York 10004, USA

Mr. R. Sagel Ing
  dS+V hoofdafdeling Bouw-en Woningtoezicht, Rotterdam, The Netherlands

Mr. Brian Smith FREng, FICE, FIStructE, FASCE
  Consultant to Flint & Neill Partnership, 21 Dartmouth Street, London, SW1H 9BP

Mr. Stefan Tietz BScEng, FREng, FICE, FIStructE
  former Senior Partner, S.B.Tietz & Partners, Consultant to Cadogan Tietz, 1 Halsey Street, London, SW3 2QH

Professor J.N.J.A. Vambersky MSc (CEng), ONRI (FIDIC)
  Delft University of Technology, CORSMIT Consulting Engineers, The Netherlands

Professor Jonathan G.M. Wood PhD, CEng, MICE, FIStructE, FIAgrE,
  Northbridge House, Chiddingfold, Surrey, GU8 4UU

Ms. Dana Wordes PE, MASCE, admitted to bar in New York and New Jersey
  Postner & Rubin Attorneys at Law, 17 Battery Place, New York, New York 10004, USA

1 Forensic engineering – the perspective from N. America

Kenneth L. Carper

This book is a compilation of experiences encountered by investigators studying failed engineering projects throughout the world. It should be of interest not only to forensic investigators, but also to practising design and construction professionals as they seek to improve their practices by learning from the experience of others, thereby avoiding failures of projects under their own direction or control.

The idea of learning from failures is not new. This, indeed, is the tradition of engineering design. Most current practices and design standards have evolved through the collective trial-and-error experience of engineering practitioners (Petroski, 1985; Feld and Carper, 1997). Open dissemination of the information gained from studying failures is essential if similar failures are to be avoided in the future. This is the purpose underlying the publication of information about failures. It is not to criticize the unfortunate individuals directly involved with the incident, but rather so that we can learn the important lessons. We cannot afford to make all the mistakes ourselves.

Forensic engineers are the pathologists of the engineering professions. The results of their investigations, when properly communicated to practitioners, can contribute to improvements in engineering design in the same way that medical pathologists have contributed to improved medical practices. In the past, carefully conducted technical investigations led to new understandings of the properties of materials (fracture mechanics) and the characteristics of foundation behavior (soil mechanics). These are but two of the many engineering fields that have their roots in failure analysis. At the present time, forensic engineers are contributing to new understandings about the human error components of failures. Many current failures, including some very costly and prominent incidents, are the results of flawed management and project delivery systems rather than technical considerations. One such failure was the 1981 collapse of the Hyatt Regency Hotel walkways in Kansas City, Missouri – a tragic event that claimed 114 lives (Fig. 1.1), and there are many other examples of failures involving human error. Management problems in such projects deserve much more discussion in the engineering literature; there are many lessons yet to be learned (Gillum, 2000; Luth, 2000; Moncarz and Taylor, 2000; Pfatteicher, 2000).

Forensic engineering is often identified with litigation. The forensic engineer may be called upon to present technical information in the courtroom and render opinions

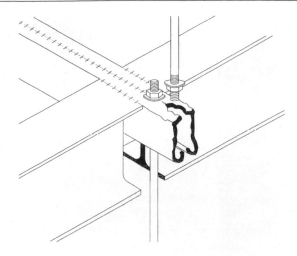

Figure 1.1   *Sketch of the Hyatt walkway failure, July 17, 1981, Kansas City, Missouri (adapted from Feld and Carper, 1997).*

on responsibility for failures. The ethical presentation of such rational information in the often emotionally charged dispute resolution arena is a very important contribution. Milton F. Lunch, former General Counsel to the U.S. National Society of Professional Engineers (NSPE), defined the field in this way:

> *Forensic Engineering is the application of the art and science of engineering in the jurisprudence system, requiring the services of legally qualified professional engineers. Forensic engineering may include investigation of the physical causes of accidents and other sources of claims and litigation, preparation of engineering reports, testimony at hearings and trials in administrative or judicial proceedings, and the rendition of advisory opinions to assist the resolution of disputes affecting life or property.*

Because failures are nearly always accompanied by controversy, the forensic engineer's work is usually overshadowed by the threat of litigation. Investigations are generally conducted with that possibility in mind. However, the contemporary role of the forensic engineer extends far beyond the legal arena. Many times, the assignment is simply to discover the source of an engineering problem and to recommend repair procedures. Accident reconstruction may be performed simply for the purpose of avoiding similar accidents in the future. Despite common perceptions, it is actually quite infrequent that a construction dispute is carried all the way to the courtroom, even in the litigious United States. Several prominent forensic engineers working in the construction industry have estimated that less than five percent of the cases they investigate are resolved through traditional litigation. Indeed, the competent forensic consultant can often be instrumental in contributing to agreements outside the court-

Figure 1.2 *Alfred P. Murrah Federal Building, Oklahoma City, Oklahoma, 1995. An act of domestic terrorism caused the loss of 168 lives (Courtesy W.G. Corley and the Federal Emergency Management Agency).*

room, by way of structured negotiations or other alternative dispute resolution techniques (Carper, 2000*a*; Ratay; 2000).

Modern society has become increasingly vulnerable to accidents and failures. Forensic engineers worldwide are helping to identify trends that affect public safety and common concerns that need attention (Carper, 1998). These topics include natural hazard mitigation, safety during construction, management and procedural issues, potential misuse of computer software, and general improvements in the area of redundancy, integrity and 'robustness' of structures. One topic of considerable discussion in the United States is the rising threat of domestic terrorism (Fig. 1.2). Teams of forensic engineers are providing valuable information that can mitigate against this unfortunate development (Corley *et al.*, 1998). Forensic engineers are finding ways to enhance the dissemination of failure-related information and are creating resources for use in undergraduate, graduate and professional education (Carper, 2000*b*). All of these activities lie outside the litigation arena, where forensic engineers are making substantial contributions to the reduction of the frequency and severity of failures.

In the United States, two organizations have played major roles in establishing the professional discipline of forensic engineering. The National Academy of Forensic

Engineers (NAFE) is affiliated with the National Society of Professional Engineers. It is a broad-based organization encompassing engineers from a wide variety of disciplines. Members of NAFE specialize in product liability engineering, traffic accident reconstruction, fire investigation, etc. In the discipline of civil engineering, the American Society of Civil Engineers has taken the lead role through its Technical Council on Forensic Engineering (ASCE/TCFE). The various committees of the TCFE are contributing in many significant areas. The TCFE works to encourage ethical practices in forensic engineering, and publishes guidelines for practising forensic investigators (ASCE/TCFE, 1989). Enhancing the dissemination of failure information throughout the industry is a major goal of the TCFE as is the preparation of related resources for education (Shepherd and Frost, 1995). The TCFE publishes the *Journal of Performance of Constructed Facilities*, a quarterly refereed journal on the causes and costs of failures. The journal contains articles on failure case studies from which valuable lessons can be learned.

Those who work in the field of forensic engineering find their assignments to be interesting, but demanding. Commitment to high standards of ethical conduct is essential if justice and fairness are to be preserved. The professional integrity of the forensic engineer is often challenged in the legal arena, and his or her technical competency is also regularly called into question. Nevertheless, the practising forensic engineer has the opportunity to bring rational information into the courtroom or other dispute resolution forum, and can serve the public by explaining the technical and procedural causes of failures in understandable language so that appropriate judicial decisions can be made.

The competent and articulate forensic engineer, during the course of an investigation, can help to restore public confidence in the engineering professions. Furthermore, opportunities sometimes arise for the dissemination of valuable information that extends far beyond the specific case involved in the investigation. This may be the most rewarding aspect of the forensic engineer's work—the opportunity to contribute to the eventual reduction of failures and accidents.

It is entirely fitting that this ambitious volume should be published in the United Kingdom, where there is an impressive history of forensic engineering. The study of engineering failures with the goal of improving design practices is a longstanding tradition in the UK. Contributions by members of the Institution of Civil Engineers (ICE) and the Institution of Structural Engineers (IStructE) in this field are known throughout the world. Indeed, the aforementioned current activities sponsored by the American Society of Civil Engineers parallel many ongoing projects within ICE and IStructE. For example, the regular reports issued by the ICE/IStructE Standing Committee on Structural Safety (SCOSS) contain reference to remarkably similar concerns that are under study by the ASCE Technical Council on Forensic Engineering (SCOSS 1997).

In 1856, while serving as President of the Institution of Civil Engineers, Robert Stevenson addressed a meeting of the ICE with these observations:

Nothing is so instructive to the younger members of the profession as the record of accidents in large works, and of the means employed in repairing the damage.

A faithful account of those accidents, and of the means by which the consequences were met, is really more valuable than a description of the most successful works. Older engineers have derived their most useful store of experience from the observation of those casualties which have occurred to their own and to other works, and it is most important that they should be faithfully recorded in the archives of the Institution.(Hammond, 1956)

Advances in the art and science of engineering have often come by way of thoughtful observation of failures. Today, there exists the unprecedented opportunity to coordinate the collection of failure-related information on an international and interdisciplinary scale. Forensic engineers throughout the world are discovering that modern societies share many common problems in the design, construction and manufacturing of engineered facilities and products. Forensic engineers are working together in an effort to develop coordinated mitigating strategies, including enhanced dissemination of information. This book, with its long list of distinguished international contributors, recognizes the synergies inherent in such cooperation. It is hoped that the reader will find this effort to be a worthy contribution to the time-honored engineering tradition of learning from failures.

References

ASCE/TCFE 1989. *Guidelines for Failure Investigation*, American Society of Civil Engineers, Technical Council on Forensic Engineering, New York, NY.

Carper, K.L. 1998. Current Structural Safety Topics in North America, *The Structural Engineer*, **76**, No. 12, (16 June).

Carper, K.L., ed. 2000a. *Forensic Engineering*, 2nd edn, CRC Press, LLC, Boca Raton, FL. (First edition was published by Elsevier Science Publishing Co., Inc., New York, NY, 1989.)

Carper, K.L. 2000b. Lessons from Failures: Case Studies as an Integral Component of the Civil Engineering Curriculum, in *Civil & Structural Engineering Education in the 21st Century*, Southampton, UK; 26–28 April 2000.

Corley, W.G., Mlakar, P., Sozen, M. and Thornton, C. 1998. The Oklahoma City Bombing: Summary and Recommendations for Multihazard Mitigation *Journal of Performance of Constructed Facilities*, American Society of Civil Engineers, New York, NY, (August).

Feld, J. and Carper, K.L. 1997. *Construction Failure,* 2nd edn, John Wiley & Sons, Inc., New York, NY.

Gillum, J.D. 2000. The Engineer of Record and Design Responsibility, *Journal of Performance of Constructed Facilities*, American Society of Civil Engineers, Reston, VA, (May).

Hammond, R. 1956. *Engineering Structural Failures*, Odhams Press, Ltd., London, UK.

Luth, G.P. 2000. Chronology and Context of the Hyatt Regency Collapse, *Journal of Performance of Constructed Facilities*, American Society of Civil Engineers, Reston, VA, (May).

Moncarz, P.D. and Taylor, R. 2000. Engineering Process Failure—Hyatt Walkway Collapse, *Journal of Performance of Constructed Facilities*, American Society of Civil Engineers, Reston, VA, (May).

Petroski, H. 1985. *To Engineer is Human*, St. Martin's Press, Inc., New York, NY.

Pfatteicher, S. 2000. The 'Hyatt Horror': Failure and Responsibility in American Engineering,

Journal of Performance of Constructed Facilities, American Society of Civil Engineers, Reston, VA, (May).

Ratay, R.T. 2000. *Forensic Structural Engineering Handbook*, McGraw-Hill Professional Publishing Co., New York, NY.

SCOSS 1997. *Structural Safety 1994-96: Review and Recommendations*, 11th Report of the Standing Committee on Structural Safety, SETO Ltd., London, UK.

Shepherd, R. and Frost, J.D. eds. 1995. *Failures in Civil Engineering: Structural, Foundation and Geoenvironmental Case Studies*, Committee on Education, Technical Council on Forensic Engineering, American Society of Civil Engineers, New York, NY.

2 Learning from history

Lawrance Hurst

Seventy-year-old subsidence

Investigation of widening horizontal cracks halfway up a basement party wall, and distortion in the front wall, led to an archive search to discover details of original construction just before and just after the First World War, and ultimately acceptance in 1944 by insurers of a claim for subsidence damage which started soon after the building was completed in 1923.

No. 2 was built for residential use in 1922/23 as an infill between two existing buildings, No. 3 on the right, built to the design of Lutyens in 1911, and No. 1 on the left, built in 1913/14. No. 2 was built as a residence on land leased from the London County Council for 99 years from 25th December 1922.

Copies of the plans showed the site dimensions and that the wall on the right is a party wall straddling the boundary. It appears from those plans as if the wall on the left may have been built as a flank wall on the site of No. 1 but became a party wall when No. 2 enclosed it, although the alterations to it shown on the original drawings of No. 2, including cutting a chase for a reinforced concrete column and cutting in flues and attaching chimney breasts, indicate that it does indeed straddle the line of junction.

It is fortunate that the architects who dealt with the 1952 alterations to make the ground, first and second floors more suitable for office use, marked those alterations on prints of the original architect's drawings. These drawings enabled the original foundation construction to be deduced. The original drawings show that No. 2 is a mixture of loadbearing brick walls and reinforced concrete floors. The attic above the top floor and the roof are timber framed. The foundation is noted as a 12" concrete raft, with 18" deep × 3'0" wide thickenings under the party walls.

The drawings clearly show, with heavy black lines, the walls which already existed either side of the site and the work that was done to them in the course of the construction of No. 2. Neither of the adjoining houses had basements and consequently it was necessary to excavate beneathe existing party walls and build new walls to enclose the lowest floor of No. 2, and to cast its reinforced concrete raft foundation.

The construction on the left shows the original wall to No. 1 built off a large reinforced concrete beam on RC columns on concrete pads, founded at a depth which scales

9 ft below the basement floor of No. 2. These RC columns are also shown and noted on the basement plan.

It was now apparent that the horizontal crack in the party wall with No. 1 about half-way up the basement storey was at the underside of the reinforced concrete beam built ten years earlier to carry the wall of No. 1. The wall below the crack is carried on the edge of the raft to No. 2 which has subsided and caused the crack. This subsidence explained the distortion in the front wall where the end, block bonded to the party wall with No. 1, had hung up leaving the rest to subside with the raft, and it also explained the cracking and distortion in the internal load-bearing partitions where the ends bonded to the party wall had hung up in the same way.

A soil investigation showed that the raft was bearing on firm alluvial soils which would have been adequate if No. 2 had been an independent building but was significantly less stiff than the ballast on which the columns supporting the party wall with No. 1 are founded.

The cause of the cracking and distortion was clearly subsidence, which was an insured risk, but would the current insurer accept liability for subsidence which started seventy years ago?

Examination of the deeds showed that the firm who had the alterations for office use carried out in 1952 were in voluntary liquidation ten years later, and the lease was assigned to the predecessors of the current insurers who passed it on to the current owners shortly before the problem was solved. They had therefore insured the building for nearly half its life and consequently accepted liability for the cost of piling through the raft to re-support it on the ballast 3 m below.

This problem had been put to several other parties whose approach was to monitor the movement and investigate the ground – very useful but meaningless without the archive research, which uncovered copies of the original architect's drawings and discovered the changes in ownership needed to deduce the cause of the cracking, develop a solution and convince insurers.

Cracks with a sense of déjà-vu

New cracking and displacement in the walls at the corners of a depository building with an internal reinforced concrete frame was found on examination to be an extension of movement which had occurred 30 years before, but which had been inactive during the intervening period.

This six-storey building was constructed in about 1907 with an internal reinforced concrete frame and solid brick loadbearing external walls. Apart from the smooth red facing bricks on the street elevations, all the brickwork is of Flettons in a strong cement mortar.

During the Second World War the upper storeys suffered bomb damage, but, since it only carried low priority, licences could not be obtained for its reinstatement until the 1950s.

The cracking at the corners of the building indicated outward movement of the walls. At the lower floors, which survived from the original construction, the movement reached a maximum of nearly 40 mm at the underside of the fourth floor upon which was superimposed a further new movement of 10 mm which monitoring showed was increasing. This new movement, showing as cracking with a fresh appearance, was found at all the floors above the fourth, which were reinstated in 1955.

The movement, taking the form of spreading at the corners, was found to be caused by the action of sulphates from the Fletton bricks on the cement in the mortar in which they are laid.

For sulphate attack to occur, water, soluble sulphates and Portland cement containing tricalcium aluminate must be present simultaneously. Soluble sulphates are present in some degree in almost all fired clay bricks and Flettons are known to be liable to give rise to sulphate problems. Following war damage, the top surfaces of the remaining walls were unprotected for some years and the pointing deteriorated and the walls became saturated, resulting in sulphate attack on the mortar. During the last 30 years, the pointing had deteriorated and moisture found its way into the body of the wall again to recommence the attack. Cement will have been used in the mortar in which the bricks are laid and the three constituents necessary for sulphate attack were therefore present.

This expansion, acting horizontally, caused the wall to lengthen and hence the ends to move out relative to the floors. It has caused vertical cracks where the body of the wall has expanded relative to the inner, dry, face and where the ends of the beams have restrained some of the expansion and, acting vertically, it has tried to lift the slabs off the beams.

This explanation for the cracking was proved correct by analysis of the mortar and was further confirmed by the lack of further movement following re-pointing of the walls using sulphate resisting cement and sealing with a suitable breathing coating.

The moral of this example is however unlikely the symptoms, look for a cause which is wholly consistent with the facts, however remote it may seem.

An office block with a sagging corner

Internal diagonal cracking and lozenging windows close to the cantilevered corner of a new, never occupied 1973 office building proved on investigation to be the symptoms of a defect which started during construction and could have concluded with a partial collapse.

This six-storey building had an unconventional structure. Thin precast reinforced concrete slabs stiffened with small ribs formed the upper floors and roof, spanned onto steel beams which were carried by precast reinforced concrete columns. This building is at the junction of two streets with the main entrance set diagonally across the corner. Continuous strips of windows above brick spandrel walls at the upper floors finished above the entrance, with full height brick panels forming a re-entrant corner above the front door.

During construction a welding specialist and engineer was called in to comment on suspected deficiencies in the fabrication of the steelwork and he was consulted when these cantilever beams over the main entrance appeared to be sagging. He checked the structural engineer's calculation for the cantilever beams and found that it was numerically correct, and he commented on a satisfactory short-term load test.

Construction continued but three years after completion the cracking and distorted windows referred to above were reported by an architect/designer brought in to try and devise ways of making the building more attractive to prospective tenants. At that stage both the calculations and the construction were critically examined and it was discovered that the cantilever beam calculation (which was numerically correct) omitted the point load on the end of the cantilever which increased the bending stress by 82%. Furthermore this bending stress was calculated on the gross section of the steel beam, taking no account of the holes in the flanges for the bolts connecting them to the precast concrete storey posts over which they cantilevered.

When the correct load was taken and the net section of the beam considered, the maximum bending stress under *dead* load was over 270 N/mm² which is 7% more than the guaranteed minimum yield stress, so the beams were relying on the slight enhancement provided by the in situ concrete casings but even so were gradually yielding. Had the building been occupied and the floors subject to imposed loads, it is possible that the beams would really have yielded and a partial collapse ensued.

The solution was to fix storey posts at the corners in the upper floors and jack in a new column in front of the ground floor entrance.

It was concluded that the building would probably not have been let even without these corner defects, because of the vast over supply of buildings in this area. This, together with the small cost of the remedial work, meant that it was not worth trying to hold the original designer to account, and the only possible legal recourse was to the local authority to try and convince them that the fault meant that the building was unoccupiable and hence empty rates paid on it should be repaid. Incidentally, the cantilever beam calculation in the building control files had been ticked!

The morals of this case are to critically examine any novel detail or system of construction and if you seek a second opinion to make sure that you ask the right person and not someone who happens to be there. The converse of the latter also applies – if you are asked a question and it is not really your subject, say so loud and clear.

The bricklayer's dilemma

The head porter of an 18-storey block of flats was accustomed to walk his dog at the end of the day and one evening on his way back he looked up at his charge and saw a noticeable bulge in the gable wall which he was sure had not been there the day before. He sensibly arranged for the area below to be cleared of cars and hoarded off so that no property was damaged and no-one was hurt when four square metres of the outer brick skin of the wall fell from seventh floor level the following morning.

This incident occurred at the time when the industry was waking up to the fact that fired clay bricks are subject to slow irreversible long-term expansion and when they are used as the outer skin of the cavity wall to a multi-storey reinforced concrete building which is subject to long term shrinkage, these opposite and inexorable movements can result in failures at the bearings of the brick skin.

This particular block of flats was built in 1962 with in situ reinforced concrete gable walls clad with an outer half-brick skin carried at each floor on a nib projecting from the concrete wall. As was the custom, the faces of the nibs were concealed by a course of thin brick slip tiles. At one end of the reinforced concrete wall was a vertical strip pierced by the windows for the bathrooms, above decorative precast concrete panels.

Ties between the brick outer skin and the concrete wall were copper butterflies cast into the concrete. They were bent into an L and the vertical part tacked to the formwork before the concrete was cast so that it could be bent down after the formwork was struck and built into the brickwork. This was also a normal and much used detail at that time, but it commonly resulted in only a proportion of the ties being actually bent down and built into the brickwork because nobody bothered to tell the bricklayers how important they were and hence if they were partially embedded in the concrete and not easy to bend down they were left.

This, however, was only partially the cause of this collapse. The main problem was at the end of the concrete wall, beside the bathroom windows and precast panels. The bricklayer had to return the outer brick skin to close the end of the cavity and naturally a vertical damp proof course (dpc) was detailed to prevent moisture transmission from the outer brick skin to the inner skin. The bricklayer's dilemma was whether to incorporate the dpc at that point or to cut it every four courses so that he could bend out the copper butterfly ties provided and tie the outer brick skin to the concrete wall. He chose the easy option, perhaps because he knew dpcs were important and perhaps because no-one had bothered to tell him how important it was, and is, for the outer skins of cavity walls to be well restrained at the ends.

The bricks expanded and the concrete frame shrank and the brick outer skin buckled outwards at its free end where the builder had omitted the ties and four square metres fell to the ground. Subsequently the brick skins to all the gables were taken down and rebuilt with adequate and sufficient ties, proper bearings and compression joints.

Failures of brick cladding to multistorey reinforced concrete buildings were widespread and were caused by extrapolating cavity wall construction from low to high rise buildings. There was a lack of appreciation of the physical properties of the materials involved and no attempt was made to tell the workforce, or even the architects, surveyors and engineers detailing the work, about the increasing need to pay attention to the detailing because of the implications of any shortfall in this respect. It is always difficult to recognise the effects and limits of extrapolation.

The London Custom House collapse

On Tuesday, 25 th January, 1825 Robert Smirke recommended that the east end of the Long Room in the London Custom House be cleared, as he feared it could collapse at any time, and it did so during the morning of the following day. This event was the culmination of a sorry chapter of events which appears to have started with the award of the contract for the new London Custom House to Peto and Miles whose tender was only 72% of the estimate of £229,000.

The new London Custom House was needed to serve the increasing trade in the Port of London, to process the documents, collect the dues and to make monthly returns to the Board of Trade. The business with the incoming and outbound sea captains, the merchants and the public was conducted in the principal chamber, termed the Long Room, a huge lofty room 199 ft long × 66 ft wide × 50 ft high.

The new Custom House was designed by David Laing, a former pupil of John Soane, with advice on the foundations from John Rennie. The site, immediately adjoining the 'disgusting spectacle' of Billingsgate, was part of the old river bed and was a confused mass of old walls, sewers, ancient quays and rubbish, through which John Rennie and David Laing made borings and discovered up to 16 to 20 ft of brick rubbish, black river mud and 'thin earth of a very unequal quality, some of it extremely soft'. It was therefore decided to use a piled foundation, with green beech piles still covered with bark, shod and hooped with iron, driven till the hammer of the engine recoiled. Beech sleepers measuring 9″ × 5″ were laid on the piles, filled in with brickwork, and a tier of beech planking was laid on these sleepers, off which the footings to the walls and piers were built.

Furthermore the wall footings were to be reinforced with a tier of oak chainbond timber, 12″ × 9″, dovetailed, halved and corked to ensure continuity.

Construction started in 1813 and the severe winter, during which there was a frost fair on the frozen river Thames, impeded the preliminary excavation and piling in what was evidently a very muddy site, criss-crossed with obstructions. The work was consequently already well behind schedule when the existing Custom House burnt down in February 1814, which increased the pressure on the contractors to complete the new building.

Eventually the Board of Commissioners were able to take up their quarters in the partially completed building in May 1817, over 15 months after the original contracted date. Practical completion was eventually achieved in November 1817, by which time the Office of Works had been involved in two investigations into unnecessary delay and alleged unprofessional conduct, and the building cost had escalated to £255,000 compared with the tender of £165,000.

In the completed building the lowest two floors of the centre block, beneath the Long Room, accommodated the King's Warehouse in which seized goods were held. These two storeys were vaulted, with an extensive series of stone and granite piers carrying two-way brick vaults, with waisted piers similar to David Alexander's vaults at Tobaco Dock.

Cracks appeared in these vaults in the spring of 1820 and Laing was naturally consulted. He claimed 'no insecurity need be apprehended therefrom' and recommended that the 'fissures be caulked and stopt with tow and oakum, and then pointed with hard cement'. Within a year the cracks had opened and the south front shed some of the lettering from the entablature. When examined, it was found to be out of plumb. Next some plaster fell from the domed ceiling of the Long Room, and by September 1823 the roof had moved far enough to reverse the fall of the gutters.

In December 1824 a 'pillar which supports one of the arches of the building' was found to have 'given way at the base in such a way as to endanger the safety of the building'. Raking and dead shores were at once erected and Robert Smirke was called in to make an independent examination. However movement continued and on 26th January 1825 part of the floor of the Long Room collapsed.

Robert Smirke shored up what was left and investigated the failure and by June

1825, when the Long Room clock was removed, dismantling was well under way, preparatory to reconstruction and remedial work.

Smirke's recommendations included reconstruction of almost the whole of the centre block off a new raft foundation and total underpinning of the east and west wings. For this foundation work he employed what was then a novel material in which he had become a specialist – lime concrete – which he had previously used for remedial work to the subsidence at Millbank Penitentiary and for the foundations for the redevelopment of the Savoy Precinct.

In the course of the underpinning and the excavation for the raft, which extended to a depth of 12 to 15 ft, the 2378 beech piles were exposed and removed, and as they were found it became apparent that they were shorter and smaller than had been specified and paid for. Peto, the contractor, had charged for 104,000 ft run of pile but had only put in 53,300 ft run, and that had an average diameter of 7½″ instead of the 9″ diameter specified.

As so often happens when buildings are taken apart in the course of reconstruction and remedial works, other defective and sub-standard work was found, such as chalk lime and stone rubbish in the spandrels of the vaulting in lieu of solid brickwork, second-hand timber in small pieces as sarking under the slates, and defective leadwork. Patent flue bricks dating from the 1820s in the chimney stacks above the west wing seen in the course of recent works indicate that Smirke's remedial work was far more extensive than had previously been reported.

These inadequacies, some of which were regarded as fraudulent, resulted in legal actions against Laing, the architect and Peto, the contractor.

The collapse and an explanation for it were reported forty years later in Edwin Nash's 1867 paper to the Royal Institute of British Architects entitled 'Remarks upon failures in construction', in which he says 'it was not occasioned by defective design or workmanship in the building itself, but by defective arrangements in preparing the foundation'. He attributes the collapse to errors in setting out the piles and says 'the entire site being incommoded with vast heaps of rubbish so it was no easy task to get all set out and executed in accurate position'. He illustrates a group of nine piles under one of the King's Warehouse piers in which the middle row was mistaken for the outer row leaving the pier overhanging the side of the group. The pier slipped off the piles 'and consequently it went down to the astonishing degree of five or six feet, burying itself to its neck, and urged downwards by the extra pressure given by the other piers and arches above, which, having lost their equilibrium, sank over towards this descending pier, and the two stories eventually fell as a disastrous ruin'.

A further item in the RIBA Transactions for 1867 explains from whence Nash obtained this information over 40 years after the event. This is the memoir of the late Charles Fowler, who had been an assistant to Laing at the time of construction of Custom House, which Laing 'left to his able and enterprising assistants', because he 'had other works in hand' and also 'was of a somewhat indolent disposition'. Amongst the remarks about the late Charles Fowler, Nash cites his act of kindness 'I wanted some information respecting the Custom House, when he not only gave me the facts I required, but also a description, and even made drawings for me, which I had not asked for'.

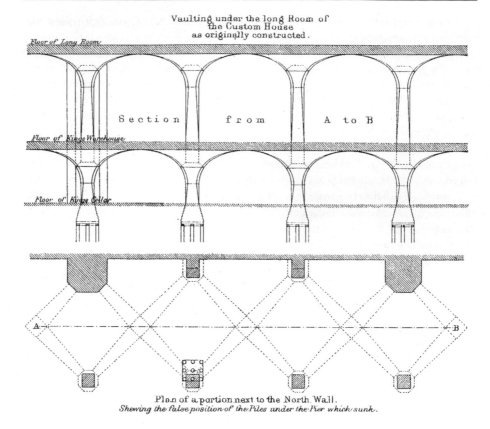

Vaulting under the long Room of
the Custom House
as originally constructed.

Floor of Long Room

Section   from   A to B

Floor of Kings Warehouse

Floor of Kings Cellar

A ——————————————————————— B

Plan of a portion next to the North Wall.
Shewing the false position of the Piles under the Pier which sunk.

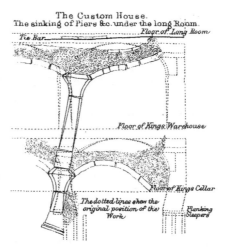

The Custom House.
The sinking of Piers &c. under the long Room.

Tie Bar ————— Floor of Long Room

Floor of Kings Warehouse

Floor of Kings Cellar

The dotted lines shew the
original position of the
Work

Planking
Sleepers

Figure 2.1 *Failure of vaulting under the long room of the Custom House (from Nash, 1867).*

This shows that the explanation for the collapse given by Nash came from one of the parties closely involved in the construction and it is interesting to consider if this is more reliable than the dry entries in the contemporary official reports. Those of us who have been directly involved in a collapse or in inadequacies in construction know that it is difficult not to give an unbiased view and whilst there is probably some truth in Fowler's explanation to Nash, it seems it is unlikely to be the whole cause.

Inaccurate setting out was cited at the time, but to a greater extent than Nash relates – Smirke reports that Laing's drawings 'represent nine piles under each of the twelve piers that supported the Long Room floor, but there are under some only four, under others three piles, and under two of the piers only two piles, although a permanent weight of upwards of 150 tons was charged upon the base of every pier'.

Pile setting out was therefore one cause, but would that have resulted in the gradual collapse which occurred, starting with cracking and distortion just over three years after the building was completed, and continued for another four years until the collapse? Whilst movement of the piers under the vaults which formed the lowest two storeys would be likely to impose extra and lateral loads on the external walls, would that cause the falls in the gutters at roof level to reverse or cause the general collapse which eventually occurred? Pile misalignment would be much less likely in the lines of piles under walls and pile failure would be expected to occur more quickly.

Furthermore, the extent of inaccuracy of piling beneath the King's Warehouse piers described in Smirke's report as quoted above would be expected to result in gross movement more quickly than the four or so years between the construction of the vaults (assuming they were structurally complete two years before practical completion) and the time when cracks were first noticed.

Buried in the reports is a much more likely cause, and it is now suggested that inadequacy and inaccuracy of piling could be a red herring. This more likely cause is decay of the tops of the piles and in the two tiers of beech forming the grillages and pile caps under the masonry. If as appears probable, the pile heads and the beech grillages were above low water, so that they were alternatively wet and dry, the conditions were ideal for the growth of wet rot and it is suggested this gradual loss of support due to increasing timber decay is more consistent with the time scale and mode of collapse than the causes cited at the time, and indeed 40 years later.

In any event, this collapse and the circumstances surrounding it provide lessons which are as valid today as they were nearly 200 years ago. These are:
- an unreasonably low tender was accepted;
- changes during construction resulted in uncontrolled bills for extras;
- an indolent and preoccupied consultant failed to identify the sub-standard construction and materials; and
- the contractor implemented measures to save himself money and increase his profit, or perhaps reduce the loss due to his inadequate tender.

Or was it simply a design fault of not having the junction where masonry took over from timber below low water, which caused the collapse and resulted in the discovery of all the other shortcomings?

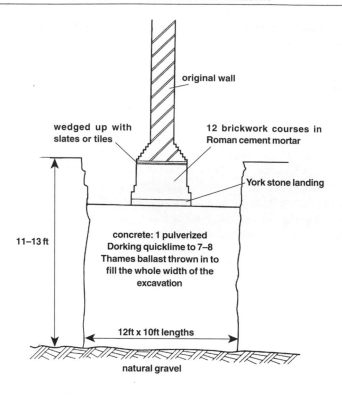

original wall

wedged up with
slates or tiles

12 brickwork courses in
Roman cement mortar

York stone landing

11–13 ft

concrete: 1 pulverized
Dorking quicklime to 7–8
Thames ballast thrown in to
fill the whole width of the
excavation

12ft x 10ft lengths

natural gravel

Figure 2.2 *Section through Robert Smirke's underpinning of Custom House (afterPasley, 1838).*

Postscript

The collapse and subsequent events did leave us however with a fascinating building. The west wing survives with the same layout and in much the same construction as originally detailed by Laing which was not changed by Smirke in the course of the remedial work he carried out. The floors are all of timber, with all the girders trussed with iron and areas of greater fire risk floored with stone slabs – an essentially Georgian large scale domestic or country house construction.

The centre block as reconstructed by Robert Smirke makes extensive use of cast iron in the columns, beams, and in the fireproof ceilings, and of wrought iron in fireproof arches in the floors as used in the King's Library at the British Museum and in a massive arch above the Long Room ceiling, hung from the timber trusses. These features are essentially early Victorian in construction style.

The west wing was hit by a high explosive bomb during the Second World War and was rebuilt with precast concrete floors and an in situ reinforced concrete frame behind reproduction facades, in 1962. This 20th century construction is, however, founded on Smirke's underpinning.

Bibliography

Colvin, H.M. (ed.) 1973. *The History of the King's Works, 1782–1851,* **VI**, pp. 422–30.

Donaldson, T.L. 1867. Memoir of the late Charles Fowler. *Trans. RIBA*, 1867–8, pp.3 & 14.

Jarvis, R.C. 1961. Laing's Custom House, 1813–27, *Transactions of the London and Middlesex Archaeological Society,* **XX**, part 4, pp. 198–213.

Mordaunt Crook, J. 1963.The Custom House scandal, *Architectural History*, VI, pp. 91–102.

Mordaunt Crook, J. 1968. Sir Robert Smirke: a pioneer of concrete construction, *Transactions of the Newcomen Society* (1965–6), **XXXVIII**, pp. 5–22.

Nash, Edwin. 1867. Remarks upon failures in construction, *Trans. RIBA*, 1867, pp. 132–3.

Pasley, C.W. 1838. *Observations on Limes, Calcareous Cements etc.,* pp.16–17, 265–9, John Weale, London.

Pasley, C.W. 1826. Practical Architecture, pp. 16, 21, 25–6, 27, 51. Lithographed notes of 1826, originally printed bythe Royal Engineers in 1862 and printed in facsimile by Donhead in 2001.

Royal Commission on the Historical Monuments of England. 1993. *The London Custom House*. RCHME.

Smith, Andrew C. 1992. Notes on research into original documents for Hurst, Peirce & Malcolm.

3   Some lessons from the past

Poul Beckmann

In the years from 1978 to 1980 a survey of some 120 structural failures was carried out in the UK under the joint aegis of the Building Research Establishment and the Construction Industry Research and Information Association. The methods and results were presented at a symposium hosted by the Institution of Structural Engineers in April 1980.

The methods used for the collection and processing of the information were novel at the time and have not, to the author's knowledge, been employed to any extent since. The results in general do however not appear to have been seriously challenged by subsequent experience and for this reason it may be worth re-capitulating some of the essential features of information gathering and some of the significant findings.

The features of the BRE/CIRIA method which were thought to distinguish it from existing methods were:

- The method of survey was devised by a panel of professional engineers with considerable experience in design and construction practice and not only in investigating cases of failure.
- The survey information was provided by professional engineers with direct and detailed knowledge of the background to the cases they reported and of the circumstances in which the failure occurred.
- The information was provided on a strictly confidential basis and was processed in a non-attributable form, thus enabling the engineers to make available all relevant information even though it may have been sensitive to claims and legal disputes. Such information would otherwise never have seen the light of day and its omission could have distorted the interpretation of the remainder of the information.
- Cases of loss of serviceability were to be reported as well as collapses. It was believed that in the past only a small proportion of cases of loss of serviceability had been studied with the result that comparative analysis was difficult or impossible.

The BRE/CIRIA questionnaire was so devised that opportunity was given to report

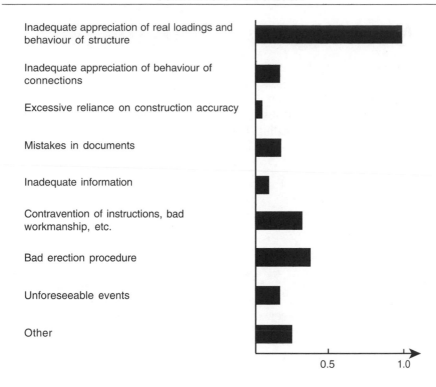

Inadequate appreciation of real loadings and behaviour of structure	
Inadequate appreciation of behaviour of connections	
Excessive reliance on construction accuracy	
Mistakes in documents	
Inadequate information	
Contravention of instructions, bad workmanship, etc.	
Bad erection procedure	
Unforeseeable events	
Other	

Figure 3.1 *Prime causes of failure in 120 case histories.*

all the known circumstances leading to and surrounding the failure as well as the mode and mechanism of the actual failure. It would also work as a checklist to ensure that all relevant available information was included by the reporter.

The opportunity was, however, also given to the reporters to supply information that was considered relevant but was not covered by the questionnaire. This avoided the possible distortion of the facts to fit a standard format. The information was coded in such a way that it could be processed, stored, retrieved and put to use.

The 120 case histories did only constitute a very small sample and the composition of the survey panel – six middle-aged consulting engineers – may have slightly biased some of the results; nevertheless, some clear trends were shown up by the statistical analysis of the results, carried out by Dr. A.C. Walker (Department of Civil Engineering, University College London) particularly the weighted values of the prime causes (Fig. 3.1) (Institution of Structural Engineers, 1980).

Some of the trends identified in the study will be illustrated by the following case histories.

Case history A

Three circular concrete tanks 35.5 m in diameter were to be covered by aluminium domes. They were situated in the UK approximately 250 m above sea level.

Tenders for design and construction were invited and the design loads were specified as wind load in accordance with British Standard Code of Practice CP3, Chap. V (1952) 'Loading' and 1.8 kN/m² snow loading. Following enquiry from one tenderer, the snow load requirement was reduced to conform to CP3, Chapter V, i.e. 0.75 kN/m².

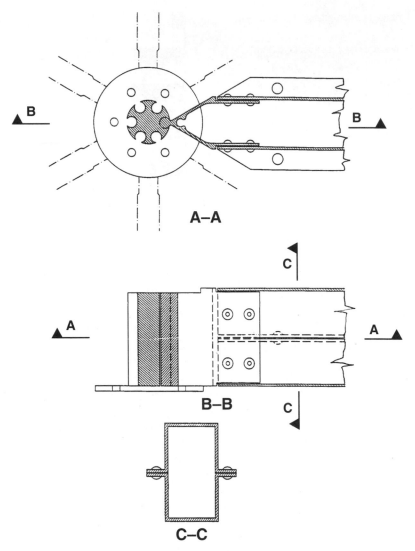

Figure 3.2 *Node and rib geometry (Case history A).*

The structure of the accepted tender was spherical with a rise of 4.1 m and consisted of a triangulated lattice of box section members, 'pop'-riveted to special end pieces which were joined together in 'hubs' at each node (Fig. 3.2). The cladding was of flat aluminium sheeting on aluminium rails.

The domes failed one after the other at 2 hour intervals after a heavy fall of wet snow which was drifting in a 13 m/sec wind.

After the failure of the first dome, attempts were made to clear the snow from the remaining ones, and in the process the maximum depth of snow was assessed to be 0.75 m and subsequent weighing of a cube, cut from the snow, gave a density of approximately 210 kg/m^3. The snow cover was seen to decrease gradually towards the meridian and there was no snow on the windward half.

The failures took the form of collapse of the leeward parts of the domes, leaving the windward halves standing. No evidence was found of failure having initiated in the box members themselves, but a large number of the end pieces were seen to be ruptured.

On investigation, the fabricator's design was found to have been based on a computer analysis, carried out at a university department. This analysis did however not include the load case for snow on part of the dome only, presumably because the code of practice at the time did not stipulate this load case. Proper appreciation of the load-carrying action of arches and domes should however have alerted the design team to this problem.

Alternative computer analyses, which were carried out as part of the failure investigation, confirmed that one-sided snow load produced member forces substantially larger than overall snow load of the same intensity. Nevertheless the lattice members were found by calculation to have adequate strength.

The geometry of the connections between the lattice members and the end pieces, joining the members to the node hubs, was such as to cause very complex stress conditions, which were not amenable to calculation by the means available at the time. As failure appeared to have initiated at these connections and as no other reliable information on their strength was available, load tests were carried out on undamaged connections, taken from the parts of the domes that had remained intact.

The average of three tests in axial compression gave a failure load of 59.1 kN which, when compared with the maximum calculated forces under different load conditions, gives the factors of safety tabulated below:

	Total load on dome	Max. compr. member force	Factor of safety
Own weight + 0.75 kN/m^2 snow on entire dome	890 kN	30.4 kN	1.93
Own weight + 0.75 kN/m^2 snow on leeward half only	519 kN	44.8 kN	1.32
Own weight + snow on leeward half, parabolically distributed with 1.6 kNm^2 maximum, as measured on site	1083 kN	69.7 kN	0.85

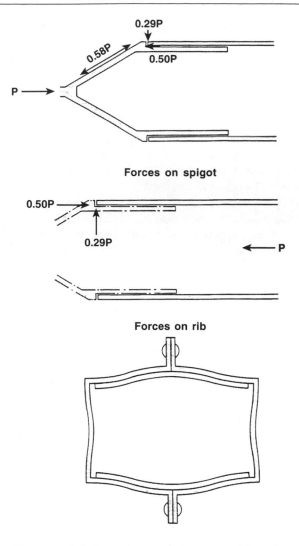

Figure 3.3 *Forces and deformations at spigot joint (Case history A).*

Had the snow loads been taken as 20% higher than in Chapter V (i.e. 0.9 in lieu of 0.75 kN/m), then the domes should have survived, but only with a factor of safety of 1.0. There is however no guarantee that the snow load (*see* Fig. 3.4), which brought about the collapse, was the heaviest to be expected in the locality, nor is it certain that the average failure load deduced from tests on three specimens represents a 'characteristic' value, let alone a minimum strength.

Had the connections, however, been designed with the then current factor of safety of 2.0 for the forces arising from the Chapter V snow load applied to one half of the

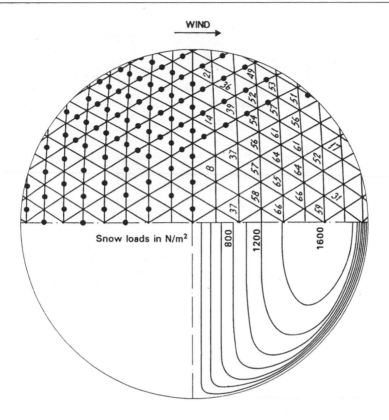

Figure 3.4 *Rib layout and snow loads (numbers in italics indicate forces in kN in compression arch; infilled circles indicate ribs in tension).*

dome only, the check calculations showed that the domes would have carried the snow load, estimated to actually have occurred, with a factor of safety of 1.25.

The main causes of the failure were judged to be:

- Failure to analyse the structure for uneven snow load,
- Failure to properly establish the strength of connections.

Contributory causes of the collapse were considered to be:

- The acceptance of CP3 Chapter V snow load in a location where heavier snow loads could be expected.
- The split responsibility between design and analysis.

Case history B

The floor structure for a building consisted of precast concrete ribbed slabs, supported

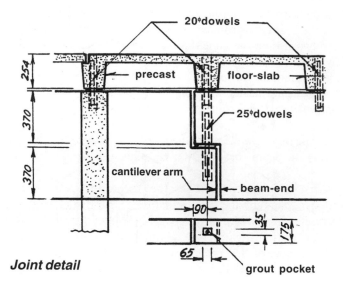

Joint detail

grout pocket

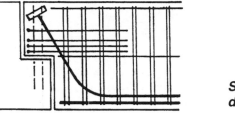

Schematic reinforcement detail

Figure 3.5 *Joint and schematic reinforcement details.*

on rectangular precast concrete beams which in turn were supported, through halved joints, on cantilever brackets having the same cross-section as the beams. Vertical dowels located the ribbed slabs to the beams and the beams to the cantilever brackets (Fig. 3.5).

The beam section had, possibly for architectural reasons, been chosen so narrow that horizontal loops to anchor the heavy main reinforcement could not be accommodated, and welded anchor blocks were used on the inclined bars in the corbels, both on the beams and on the cantilever arms.

Prior to erection frost damage occurred to some of the dowel pockets which had filled up with rain whilst the units were stored in the open. Where visible cracks were found, the corbels were cut off and re-concreted.

When the heavy floor finishes had been applied after erection, creep deflection took place and the beam ends rotated. This rotation was restrained by the dowels which connected two ribs on the same floor slab to the beam and to the cantilever bracket respectively.

The restraint of the dowels against the end rotation was only resisted by relatively light reinforcement (the half-joints being designed to act as inclined ties of heavy reinforcement with horizontal concrete 'struts'), and cracks developed over a period of two to three years.

On investigation it was found that in many instances the reinforcement had been misplaced in the beam corbels, and where this was the case, the cracking was more severe.

A design check revealed that whilst the reinforcement was adequate for the normal vertical design loads and the horizontal forces arising from wind loads etc., it did not have any reserve capacity for the forces arising out of the combined effects of residual shrinkage and restraint against the end rotation caused by creep deflection.

The direct cause of the cracking was the failure to allow in the design for the end rotation caused by creep deflection. This was inextricably bound up with the choice of a narrow beam section which limited the reinforcement loops round the dowel holes to small diameter bars which were easily displaced. Whilst no conclusive evidence could be produced, there was a strong indication that undetected frost cracking at the dowel pockets contributed to the large number of corbels which subsequently cracked.

The cracking could have been avoided, or at least minimised by:
- the choice of a wider beam section to accommodate horizontal loops of heavy reinforcing bars round the dowels. This would allow the corbels to be designed to act with inclined concrete struts and horizontal ties in the shape of heavy horizontal loops. These would have been able to resist the restraint forces from the dowels and be less likely to be displaced.
- more careful storage to avoid water filling the grout pockets and causing frost damage.

Less emphasis on calculations by checking authorities and greater attention to constructional details might also have created a climate in which the engineer would not have based his structural sizes on initial calculations without checking that adequate reinforcement, properly detailed, could be accommodated.

Case history C

The roof of a building was constructed of prestressed, precast concrete beams supporting precast concrete slabs acting as permanent formwork for an in-situ concrete topping overlaid by aerated screed and asphalt (*see* Fig.3.6). The prestressed beams were spanning 12.6 m between an in-situ reinforced concrete spine beam and a series of precast concrete edge beams.

The edge beams were 0.92 m deep and 0.15 m wide. The pre-tensioned roof beams were of I-profile 0.34 m deep overall with a top flange 0.58 m wide and bottom flange 0.30 m wide; they were spaced at 1.57 m ctrs. Their 'inboard' ends were cast into the in-situ spine beam whilst the 'outboard' ends were resting in 50 mm deep pockets cast into the vertical inside face of the edge beams.

In addition to the seating pockets for the prestressed beams, the edge beams were also recessed to accommodate a blind box at their bottom edge. These encroachments on the already narrow section led to the adoption of a reinforcement arrangement which had no main steel under the seating pockets for the prestressed beams. Bearing stresses on the shallow seating and vertical shear stresses under the pockets had been calculated to be within the permissible values in the code of practice, current at the time (CP114). Only a 12.7 mm 'inverted hat' bar was provided under the seatings, and this was detailed in a way that made it difficult to maintain in position, as the edge beams were cast with the outside face down, so as to produce a smooth fascia.

Stirrups projected from the top flange of the pre-tensioned beams into the in-situ topping and the edge beams had 9.5 mm mild steel bars at 300 mm centres projecting horizontally into the in-situ topping, thus providing some tying together, but not at the level of the seating.

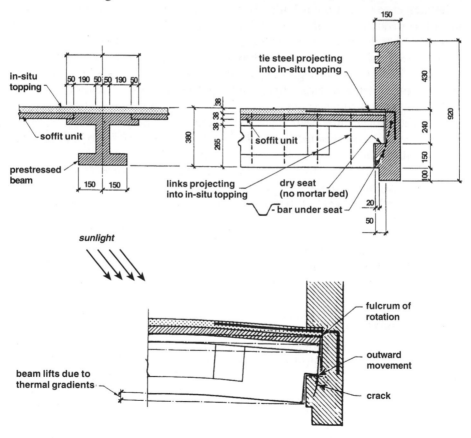

Figure 3.6 *Structural arrangement and mechansim of failure (Case history C).*

To give the fascia a dark colour the edge beams were made with a dark crushed rock aggregate and high alumina cement (HAC). (The particular aggregate was subsequently found to be highly alkaline). High alumina cement was also used for the pre-tensioned roof beams.

On a June morning, following a period of cool nights and very hot sunny days, a portion of the roof, measuring approximately 11.3 m by 5.6 m collapsed bodily on to the floor below. Three of the pre-tensioned beams under the collapsed part of the roof had first punched through their seatings in the edge beams and/or slid off them, and had subsequently broken out of their encasement in the in-situ spine beam. A fourth roof beam had pulled the bottom off its seating pocket but had remained wedged against the edge beam, and was propped soon after.

The ends of the pre-tensioned beams were largely intact and extensive core tests, UPV measurements and a subsequent test to destruction of the 'wedged' beam showed that the roof beams were structurally adequate.

Cores taken from the precast edge beams gave strengths ranging from 7–15 N/mm² and some samples crumbled during coring. Chemical analysis indicated mix proportions of 1 part of cement to 3.6 parts of fine to 1.9 parts of coarse aggregate, which corresponded to a misprint in the specification which quoted 1 cement:4 fines:2 coarse in lieu of the intended 1:2:4. This analysis also indicated a water/cement ratio around 0.65; this corresponded well with the core strengths and with results of trial castings carried out as part of the investigation, with various mix designs. These were aimed at reproducing the texture with large water blisters that had been found on the inside vertical faces of the seating pockets (cast against top shutter).

The actual mechanism which led to the failure was found to be thermal hogging of the prestressed beams. By quasi-static heat flow analysis, it was found that the temperature difference between the top of the screed and the web of the beams could vary from −18° C after a cold night, to 24° C after some hours of intense summer sunshine. The tie steel projecting from the edge beams into the in-situ topping acted as a hinge, and the end rotation of the pre-tensioned beams, accompanying the thermal flexure caused by the two extremes of temperature gradient corresponded to a horizontal movement at the level of the seating of approximately 2 mm. This movement was restrained by the friction on the seating and that restraint in turn produced a tension, perpendicular to the face of the edge beam, when the pre-tensioned roof beam hogged. The shear stress, whilst permissible according to CP114 (1957) was, according to CP110, very high for an unreinforced section. The superimposition of a direct tension could therefore easily have led to failure of even good concrete, let alone severely converted HAC concrete. Alternatively the repeated grinding movement on the seating would have caused it to crumble and form a slope for the beam end to slide down.

The high alumina cement was *not* the main cause of the collapse. In another building, investigated about the same time, a similar structural arrangement was found. The main differences were the use of Portland cement throughout and a more efficient reinforcement detail under the seating pockets. In this case the seatings were seen to have cracked in a similar way to that observed on some of those of the roof in question, adjacent to the collapse. In this second building the reinforcement had prevented the collapse.

The causes of the collapse were:
- bad reinforcement detailing arising out of inadequate depth of seating on the edge beam.
- failure to provide adequate connection, at seating level, between roof beams and edge beam, and thus failing to provide safe restraint against relative movement due to thermal hogging.

The effects were aggravated by:
- the use of high alumina cement with an excessive water/cement ratio and unsuitable mix proportions;
- the use of an aggregate that was alkaline together with high alumina cement.

Case history D

An office building, with the external walls consisting of storey-height load bearing precast concrete panels, was in the process of construction. The end bays of the floors consisted of precast, prestressed, concrete planks spanning between wall panels, forming the facade of the building, and a precast concrete spine beam that spanned between a column and the wall panels which formed the gable end. The prestressed planks were thus parallel to the gable end. Monolithic action of the floors was to be achieved by an in-situ topping with the usual mesh reinforcement, tie bars, etc. (*see* Fig. 3.7).

Erection of a floor had reached the stage when the wall panels had been placed and laterally braced by raking push/pull struts attached to the previous floor slab. The precast spine beam had been placed and levelled on the column at one end and on a seating formed in the panels of the end wall. The exact levelling was achieved by stacks of steel shims with a neoprene shim to take up the effect of unevenness of the bearing and the end rotation of the beam. The final design called for lacer bars to be placed within the overlapping loops of reinforcement at either end of the beam, but these had not been placed when erection of the floor planks commenced. The floor planks were being lowered from a tower crane by means of a 'scissors' type grab, the release of which required the planks to be spaced apart and subsequently levered into position after being released from the crane.

The last plank adjacent to the end wall, and on one side of the beam, had been lowered into position and the workmen were walking off for their tea break before adjusting the final plank, when there was a collapse of the spine beam and the floor planks bearing on it.

In the subsequent investigation the spine beam was found to be intact and to have dropped bodily at its 'outboard' end, which was now resting on the floor below, whilst the other end was still supported on the top of the column. There was a fracture of part of the nib which had been bearing on the wall panels. This fracture was completely 'clean' and ran diagonally in the opposite sense to what would have been the direction of a shear crack.

The investigation initially considered the following causes for the initiation of the collapse:

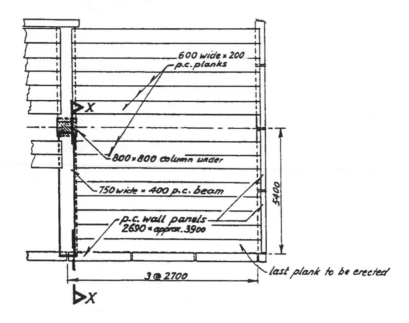

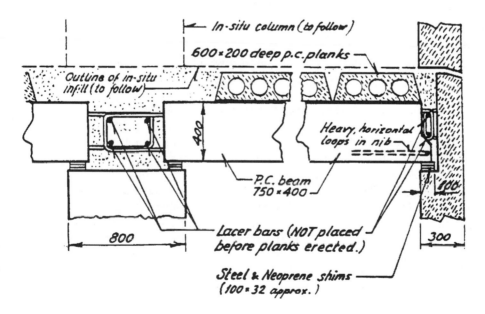

Figure 3.7 *upper: Plan of floor prior to collapse; lower: section X-X (Case history D).*

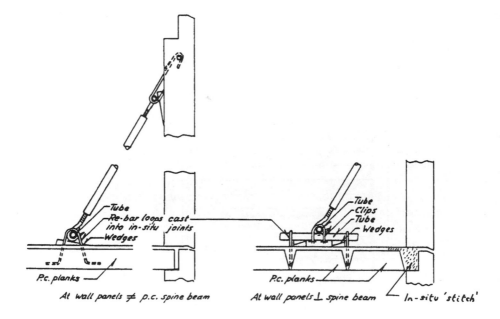

At Wall panels ≠ p.c. spine beam   At wall panels ⊥ spine beam   In-situ 'stitch'

Figure 3.8 *Details of fixings of push-pull struts (Case history D).*

- bending and/or shear failure of floor planks and/or spine beam;
- tilting of the spine beam due to one-sided loading; and
- nib failure of the spine beam due to the support reaction being concentrated by the beam resting only on the very small areas of the steel shims.

The physical evidence ruled out all of these causes.

It was deduced, by elimination, that the wall panels supporting the outer end of the spine beam must have been wedged out by one of two of mechanisms:

- the prestressed floor planks were very springy prior to the application of the in-situ topping. If the last plank had been left by the crane with its edge just overlapping on the inside of the end wall panels and one of the workmen had walked on it, it is possible that this would have moved the wall panels out and made the stacks of shims unstable, following which the outer end of the spine beam would drop and the floor planks with it.
- any attempt to lever the last plank into position with a crowbar could have forced the end wall panel outwards, thus depriving the beam of its support.

In support for these two hypotheses was the fact that the lower fixing of the push/pull struts for the end wall fixing did not have as positive a horizontal fixing as the push/pull struts for the side wall panels (*see* Fig. 3.8).

The collapse would have been prevented if the lacer bars had been placed through the overlapping reinforcement loops before the floor planks were lowered on to the beam and a more positive horizontal fixing for the push/pull struts would also have helped.

What are the lessons to be learnt from these case histories and how do they relate to the overall findings of the BRE/CIRIA survey?

- Inadequate understanding of the real behaviour of the structure played a major role in the first three cases. In case A it was the response of the structure to a probable configuration of primary load that was overlooked; in cases B and C it was what would normally be considered secondary effects, that were primarily responsible for the failures.
- Inadequate appreciation of the behaviour of the connections, and hence poor detailing of these, contributed to the failures in cases A, B and C.
- Excessive reliance on construction accuracy contributed in cases B and C.
- There was a gross mistake in the documents in case C, but in the event it did not make much difference.
- There appears not to have been any significant lack of information in any of the four cases, nor any gross contravention of instructions (although the uncritical acceptance by the contractor of the blatantly wrong mix proportions in case C borders on negligence). Bad workmanship contributed to the failure in case B and there was a strong element of it in case D.
- Case D was clearly one of bad erection procedure.

Unforeseeable events and 'other' causes were not represented in these cases. Common to all four cases was the fact that the conventional design calculations had shown that the designs satisfied the requirements of the codes of practice current at the time.

Reference

Institution of Structural Engineers 1980. *Structural Failiures in Buildings.* Papers presented at a symposium held at the Café Royal, London, 30 April, 1980 (reprinted 1981, 1984).

4 Propping up Pisa

John Burland

Introduction

In 1989 the civic tower of Pavia collapsed without warning, killing four people. The Italian Minister of Public Buildings and Works appointed a commission to advise on the stability of the Pisa tower. The commission recommended closure of the tower to the general public and this was instituted at the beginning of 1990. There was an immediate outcry by the Mayor and citizens of Pisa who foresaw the damage that the closure would inflict on the economy of Pisa, heavily dependent on tourism as it is. In March 1990 the Prime Minister of Italy set up a new commission, under the chairmanship of Professor Michele Jamiolkowski, to develop and implement measures for stabilising the tower. It is the fifteenth commission this century and its membership covers a number of disciplines including structural and geotechnical engineering, architecture, architectural history, archaeology and restoration.

It is not widely appreciated that the decree establishing the commission has never been ratified. In Italian law a decree has to be ratified by the Italian Parliament within two months of publication or else it falls. Thus, every two *months*, the commission's decree has to be renewed and on a number of occasions the work has been suspended because of delays in renewal. Such an arrangement makes the commission very vulnerable to media and political pressures and long-term planning is very difficult.

Details of the tower and ground profile

Figure 4.1 shows a cross-section through the 14,500 t tower. It is nearly 60 m high and the foundations are 19.6 m in diameter. At present the foundations are inclined due south at about 5.5° to the horizontal. The average inclination of the axis of the tower is somewhat less due to its slight curvature as will be discussed later. The seventh cornice overhangs the first cornice by about 4.5 m. Construction is in the form of a hollow cylinder. The inner and outer surfaces are faced with marble and the annulus between these facings is filled with rubble and mortar within which extensive voids have been found. The spiral staircase winds up within the annulus.

Figure 4.2 shows the ground profile underlying the tower. It consists of three distinct layers. Layer A is about 10 m thick and primarily consists of estuarine deposits laid down under tidal conditions. As a consequence, the soil types consist of rather variable sandy and clayey silts. At the bottom of layer A is a 2 m thick medium dense fine sand layer (the upper sand). Based on sample descriptions and cone tests, the material to the south of the tower appears to be more clayey than to the north and the sand layer is locally much thinner. Therefore, to the south layer A could be expected to be slightly more compressible than to the north. Layer B consists of soft sensitive normally consolidated marine clay which extends to a depth of about 40 m. The upper clay, known as the Pancone Clay, is very sensitive to disturbance which causes it to lose much strength. The lower clay is separated from the Pancone Clay by a sand layer (the intermediate sand) overlain by a layer of stiffer clay (the intermediate clay). The Pan-

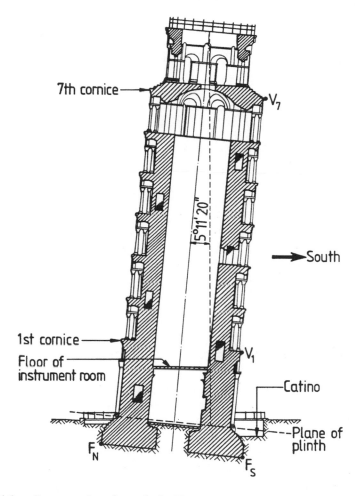

Figure 4.1 *Cross-section through the Tower.*

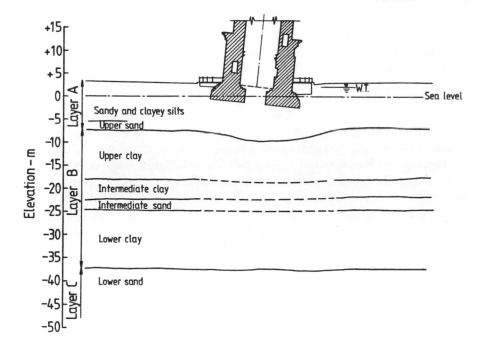

Figure 4.2 *Soil profile beneath the Tower.*

cone Clay is laterally very uniform in the vicinity of the tower. Layer C is a dense sand which extends to considerable depth (the lower sand).

The water table in layer A is between 1 m and 2 m below ground surface. Pumping from the lower sand has resulted in downward seepage from layer A with a vertical pore pressure distribution through layer B which is slightly below hydrostatic.

The many borings beneath and around the tower show that the surface of the Pancone Clay is dished beneath the tower from which it can be deduced that the average settlement is 2.5–3.0 m.

History of construction

The tower is a campanile for the cathedral, construction of which began in the latter half of the 11th century. Work on the tower began on 9th August, 1173 (by the modern calendar). By about 1178 construction had progressed to about one quarter of the way up the fourth storey when work stopped. The reason for the stoppage is not known but had it continued much further the foundations would have experienced a bearing capacity failure within the Pancone Clay. The work recommenced in about 1272 by which time the strength of the clay had increased due to consolidation under the weight of the tower. By about 1278 construction had reached the seventh cornice when work again stopped, possibly due to military action. Once again, there can be no doubt that, had work continued, the tower would have fallen over. In about 1360 work on the bell chamber was commenced and was completed in about 1370 – nearly 200 years after commencement of the work.

It is known that the tower must have been tilting to the south when work on the bell chamber was commenced as it is noticeably more vertical than the remainder of the tower. Indeed on the north side there are four steps from the seventh cornice up to the floor of the bell chamber while on the south side there are six steps. Another important historical detail is that in 1838 the architect Alessandro Della Gherardesca excavated a walk-way around the foundations. This is known as the catino and its purpose was to expose the column plinths and foundation steps for all to see as was originally intended. This activity resulted in an inrush of water on the south side, since here the excavation is below the water table, and there is evidence to suggest that the inclination of the tower increased by as much as half a degree.

History of tilting

One of the first actions of the commission was to undertake the development of a computer model of the tower and the underlying ground that could be used to assess the effectiveness of various possible remedial measures. Calibration of such a model is essential and the only means of doing this is to attempt to simulate the history of tilting during and subsequent to construction of the tower. Hence it became apparent very early on that it was necessary to learn as much as possible about the history of the tilt of the tower. In the absence of any documentary evidence all the clues to the history of tilt lie in the adjustments made to the masonry layers during construction and in the shape of the axis of the tower.

Over the years a number of measurements of the dimensions of the tower have been made and many of them are conflicting. The Polvani Commission measured the thickness of each of the masonry layers and its variation around the tower (Ministero dei Lavori Pubblici, 1971). This information has proved extremely valuable in unravelling the history of tilt.

Figure 4.3 shows the shape of the axis of the tower deduced from the measured relative inclinations of the masonry layers assuming that construction proceeded perpendicular to each masonry layer. This shape compares favourably with other independent measurements at a few locations up the tower. It can be seen that the axis is curved. For years the tower has been unkindly referred to as having a banana shape. I prefer to call it a question mark (?) so as to reflect the enigma of the tower. Some important observations can be made from the measurements on the masonry layers. For most of the storeys, construction took place using parallel sided blocks of masonry. With one or two notable exceptions adjustments only took place close to each floor using tapered blocks. The most important exception can be seen in Fig. 4.3 where there is an obvious kink one quarter the way up the fourth storey. It will be recalled that the construction remained at this level for about 100 years. Evidently the tower was tilting significantly when work recommenced and the masons made adjustments to correct it.

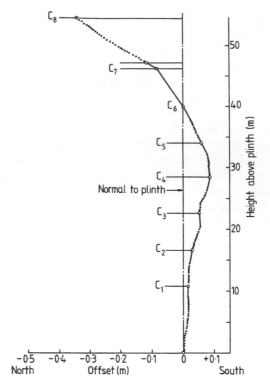

Figure 4.3 *Shape of the tower's axis deduced from the relative inclination of masonry layers.*

We see that the history of the tilting of the tower is tantalisingly frozen into the masonry layers. If only we knew the rules that the masons followed in adjusting for the tilt we would be able to unravel the history. We have to put ourselves in the place of a mason or architect in the 12th or 13th century and ask: "What is the most practical thing to do when you arrive at a given floor and find that the tower is out-of-plumb"?

A widely accepted hypothesis is that the masons would always try to keep the masonry layers horizontal and the Polvani Commission adopted this. Although this seems reasonable for a low aspect ratio building like a cathedral, it does not make sense for a tower since it would tend to perpetuate the overall out-of-plumb. After a few trials, a child building a tower of bricks on a carpet will soon learn to compensate for any tilt by attempting to place successive bricks over the centre of the base of the tower i.e. by bringing the centre of the tower back vertically over the centre of the foundations (or possibly even further, away from the direction of tilt). Therefore an alternative hypothesis is one in which the masons aimed to bring the centre line of the tower back, vertically over the centre of the foundations at the end of each storey. The architectural historians on the commission are satisfied that the masons would have had the technology to make such an adjustment, particularly as the stones for each storey were carved and assembled on the ground prior to hoisting into position.

Figure 4.4 shows the reconstructed history of inclination of the foundations of the tower using the alternative hypothesis (Burland and Viggiani, 1994). In this figure the weight of the tower at any time is plotted against the deduced inclination. It can be seen that initially the tower inclined slightly to the north amounting to about 0.2° in 1272 when construction recommenced. As construction proceeded the tower began to move towards the south at an increasing rate. In 1278, when construction had reached the seventh cornice, the tilt was about 0.6°. During the 90 year pause, the tilt increased to about 1.6°. After the completion of the bell chamber in about 1370 the inclination of the tower increased dramatically. The point dated 1817 is based on measurements made by two British architects Cressy and Taylor using a plumb line. A further measurement was made by the Frenchman Ruhault de Fleury in 1859 which showed that the excavation of the catino by Gherardesca in 1838 caused a significant increase of inclination. The history of tilting depicted in Fig. 4.4 has been used to calibrate numerical and physical models of the tower and underlying ground.

Computer modelling of the tower movements

The analysis was carried out using a suite of finite element geotechnical computer programs developed at Imperial College and known as ICFEP (Potts and Gens, 1984). The constitutive model is based on critical state concepts (Schofield and Wroth, 1968)

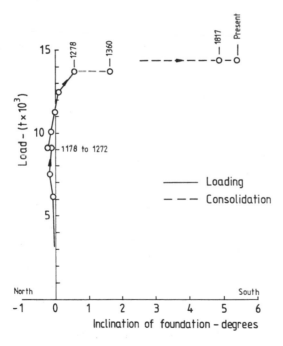

Figure 4.4 *Deduced history of the tower's inclination during and subsequent to construction.*

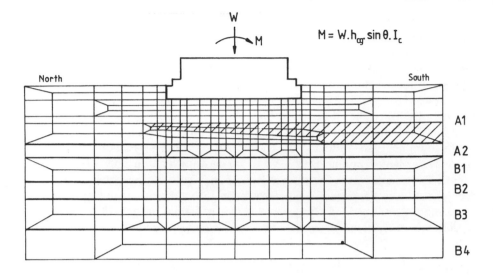

Figure 4.5 *Finite element mesh in the vicinity of the tower foundation.*

and is non-linear elastic work-hardening plastic. Fully coupled consolidation is incorporated so that time effects due to the drainage of pore water out of or into the soil skeleton are included.

It must be emphasised that the prime objective of the analysis was to develop an understanding of the mechanisms controlling the behaviour of the tower (Burland and Potts, 1994). Accordingly a plane strain approach was used for much of the work and only later was three-dimensional analysis used to explore certain detailed features.

The layers of the finite element mesh matched the soil sub-layering that had been established from numerous extensive soil exploration studies. Figure 4.5 shows the mesh in the immediate vicinity of the foundation. In layer B (*see* Fig. 4.2) the soil is assumed to be laterally homogeneous. However a tapered layer of slightly more compressible material was incorporated into the mesh for layer A1 as shown by the cross-hatched elements in Fig. 4.5. This slightly more compressible region represents the more clayey material found beneath the south side of the foundation as discussed previously. In applied mechanics terms the insertion of this slightly more compressible tapered layer may be considered to act as an 'imperfection'. The overturning moment generated by the lateral movement of the centre of gravity of the tower was incorporated into the model as a function of the inclination of the foundation as shown in Fig 4.5.

The analysis was carried out in a series of time increments in which the loads were applied to the foundation to simulate the construction history of the tower. The excavation of the catino in 1838 was also simulated in the analysis. Calibration of the model was carried out by adjusting the relationship between the overturning moment generated by the centre of gravity and the inclination of the foundation. A number of runs were carried out with successive adjustments being made until good agreement was obtained between the actual and the predicted present day value of the inclination.

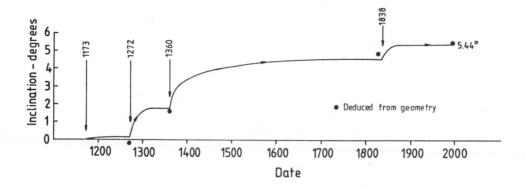

Figure 4.6 *Relationship between time and inclination for the computer simulation of the history of the Pisa Tower.*

Figure 4.6 shows a graph of the predicted changes in inclination of the tower against time, compared with the deduced historical values. It is important to appreciate that the only point that has been predetermined in the analysis is the present-day value. The model does not simulate the initial small rotation of the tower to the north. However, from about 1272 onwards there is remarkable agreement between the model and the historical inclinations. Note that it was only when the bell chamber was added in 1360 that the inclination increased dramatically. Also of considerable interest is the excavation of the catino in 1838 which results in a predicted rotation of about 0.75°. It should be noted that the final imposed inclination of the model tower is 5.44° which is slightly less than the present-day value of 5.5°. It was found that any further increase in the final inclination of the model tower resulted in instability – a clear indication that the tower is very close to falling over.

Burland and Potts (1994) concluded from a careful study of the computer model that the impending instability of the tower foundation is not due to a shear failure of the ground but can be attributed to the high compressibility of the Pancone Clay. This phenomenon was called 'leaning instability' by the late Edmund Hambly (1985) who used it to explain the lean of the tower. No matter how carefully the structure is built, once it reaches a critical height the smallest perturbation will induce leaning instability. As pointed out by Hambly, "…leaning instability is not due to lack of strength of the ground but is due to insufficient stiffness, i.e. too much settlement under load." Children building brick towers on a soft carpet will be familiar with this phenomenon!

In summary, the finite element model gives remarkable agreement with the deduced historical behaviour of the tower. It is important to emphasise that the predicted history of foundation inclinations and overturning moments were self-generated and were not imposed externally in a predetermined way. The only quantity that was used to calibrate the model was the present-day inclination. The analysis has demonstrated that the lean of the tower results from the phenomenon of settlement instability due to the high compressibility of the Pancone Clay. The role of the layer of slightly increased compressibility beneath the south side of the foundations is to act as an 'imperfection'. Its principal effect is to determine the direction of lean rather than its magnitude. The main limitation of the model is that it is a plane strain one rather than fully three-dimensional. Also, the constitutive model does not deal with creep so that no attempt has been made to model the small time-dependent rotations that have been taking place during the 20th century and which are described in the next section. Nevertheless the model provides important insights into the basic mechanisms of behaviour and has proved valuable in assessing the effectiveness of various proposed stabilisation measures. Its role in evaluating the effectiveness of the temporary counterweight solution is described in the section *Temporary stabilisation of the foundations* on page 45.

Observed behaviour of the tower this century

Change of inclination

For most of the 20th century the inclination of the tower has been increasing. The study of these movements has been important in developing an understanding of the behaviour of the tower and has profoundly influenced the decisions taken by the commission. It is important to appreciate that the magnitudes of the movements are about three orders of magnitude less than the movements that occurred during construction.

Figure 4.7 is a plan view of the Piazza dei Miracoli showing the location of the baptistry, cathedral and tower. Since 1911 the inclination of the tower has been measured regularly by means of a theodolite. The instrument is located at the station marked E and the angles between station D and the first cornice (V_1 in Fig. 4.1) and between

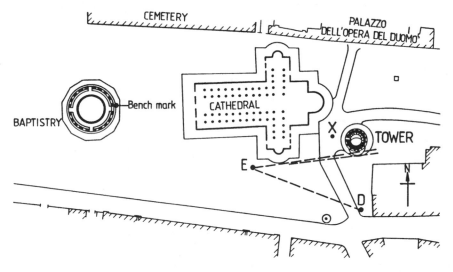

Figure 4.7 *Plan of the Piazza dei Miracoli.*

station D and the seventh cornice (V$_7$) are measured. The difference between these two angles is used to calculate the vertical offset between the seventh and first cornice and hence the inclination of the tower.

In 1928 four levelling stations were placed around the plinth level of the tower and were referred to a bench mark on the baptistry. Readings were taken in 1928 and 1929 but not again until 1965, when fifteen levelling points were installed around the tower at plinth level and about seventy surveying monuments were located around the Piazza.

In 1934 a plumb line was installed in the tower, suspended from the sixth floor and observed in an instrument room whose location is shown in Fig. 4.1. The instrument was designed by the engineers Girometti and Bonechi and is known as the GB pendulum. Also in 1934 a 4.5 m long spirit level was installed within the instrument room. The instrument rests on brackets embedded in the masonry and can be used to measure both the north-south and east-west inclination of the tower. The instrument was designed by the officials of the Genio Civile di Pisa and is known as the GC level.

Figure 4.8 shows the change of inclination with time since 1911. From 1934 to 1969 the GC level was read regularly once or twice a year except during the second world war. For some reason readings with the GC level ceased in 1969 but fortunately precision levelling on the fifteen points around the tower began in 1965 and continued regularly until 1985. In 1990 Professor Carlo Viggiani and I read the GC level again and found that the inclination agreed to within a few seconds of arc with that derived

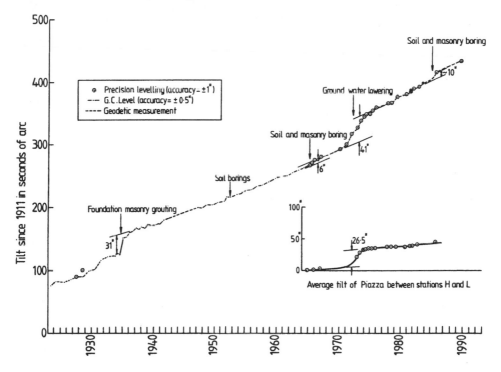

Figure 4.8 *The change in inclination of the foundations since 1911.*

from the precision levelling around the plinth.

It can be seen from Fig. 4.8 that the inclination-time relationship for the tower is not a smooth curve but contains some significant 'events'. In 1934, Girometti drilled 361 holes into the foundation masonry and injected about 80 t of grout with a view to strengthening the masonry. This activity caused a sudden increase in tilt of 31″. In 1966 some soil and masonry drilling took place and caused a small but distinct increase of tilt of about 6″. Again, in 1985 an increase in tilt of 10″ resulted from masonry boring through the foundations. In the late 1960s and early 1970s pumping from the lower sands caused subsidence and tilting towards the south-west of the Piazza. This induced a tilt of the tower of about 41″. When pumping was reduced the tilting of the tower reduced to its previous rate. It is clear from these events that the inclination of the tower is very sensitive to even the smallest ground disturbance. Hence any remedial measures should involve a minimum of such disturbance. The rate of inclination of the tower in 1990 was about 6 seconds of arc per annum or about 1.5mm at the top of the tower.

The motion of the tower foundation

Previously, studies have concentrated on the changes of inclination of the tower. Little attention has been devoted to the complete motion of the foundations relative to the surrounding ground. The theodolite and precision levelling measurements help to clarify this. It will be recalled from Fig. 4.7 that angles were measured relative to the line ED. Hence it is possible to deduce the horizontal displacements of the tower relative to point D. Figure 4.9 shows a plot of the horizontal displacement of point V_1 on the first cornice relative to point D since 1911. Also shown, for comparison, is the relative vertical displacements between the north and south sides of the foundation (points F_N and F_S). It can be seen that up to 1934 the horizontal movement of V_1 was very small. Between 1935 and 1938, following the work of Girometti, point V_1 moved southwards by about 5 mm. No further horizontal movement took place until about 1973 when a

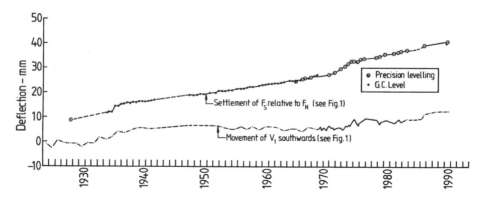

Figure 4.9 *Horizontal displacement since 1911 of V_1 on the first cornice.*

– 42 –

further southward movement of about 3 mm took place as a result of the ground water lowering. A further small horizontal movement appears to have taken place in about 1985 as a result of masonry drilling at that time. These observations reveal the surprising fact that during steady-state creep-rotation point V_1 on the first cornice does not move horizontally. Horizontal movements to the south only take place when disturbance to the underlying ground takes place.

Study of the precision levelling results shows that between 1928 and 1965 the centre of the foundations at plinth level rose by 0.3 mm relative to the baptistry – a negligible amount. Between 1965 and 1986 the relative vertical displacement between the centre of the plinth and a point a few metres away from the tower was again negligible. Thus, not only does point V_1 not move horizontally during steady-state creep, but also negligible average settlement of the foundations has taken place relative to the surrounding ground.

The observations described above can be used to define the rigid-body motion of the tower during steady-state creep-rotation as shown in Fig. 4.10. It can be seen that the tower must be rotating about a point approximately located level with point V_1 and vertically above the centre of the foundation. The direction of motion of points F_N and F_S are shown by vectors and it is clear that the foundations are moving northwards with F_N rising and F_S sinking.

Conclusions from the observed motion of the tower foundations

The discovery that the motion of the tower is as shown in Fig. 4.10 has turned out to be a most important finding in a number of respects. Previously it had been believed that the foundations were undergoing creep settlements with the south side settling more rapidly than the north. However, the observation that the north side had been steadily rising led to the suggestion that the application of load to the foundation masonry on the north side could be beneficial in reducing the overturning moment (Burland, 1990).

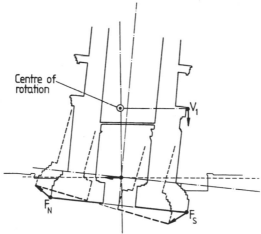

Figure 4.10 *Motion of the tower during steady-state creep-rotation.*

The form of foundation motion depicted in Fig. 4.10 leads to the very important conclusion that the seat of the continuing long-term tilting of the tower lies in layer A and not within the underlying Pancone Clay as had been widely assumed. It can therefore be concluded that this stratum must have undergone a considerable period of ageing since last experiencing significant deformation. Thus, in developing the computer model, it is reasonable to assume that the clay has an increased resistance to yield subsequent to the excavation of the catino in 1838. This conclusion has proved of great importance in the successful analysis of the effects of applying the lead counterweight (Burland and Potts, 1994).

The continuing foundation movements tend to be seasonal. Between February and August each year little change in the north-south inclination takes place. In late August or early September the tower starts to move southward and this continues until December or January, amounting to an average of about 6 arc seconds. In the light of the observed motion of the tower foundations, the most likely cause of these seasonal movements is thought to be the sharp rises in ground-water level that have been measured in layer A resulting from seasonal heavy rainstorms in the period September to December each year. Thus, continuing rotation of the foundations might be substantially reduced by controlling the water table in layer A in the vicinity of the tower.

Temporary stabilisation of the tower

There are two distinct problems that threaten the stability of the tower. The most immediate one is the strength of the masonry. It can be seen from the cross section in Fig. 4.1 that at first floor level there is a change in cross section of the walls. This gives rise to stress concentrations at the south side. In addition to this, the spiral staircase can be seen to pass through the middle of this change in cross section giving rise to a significant magnification in the stresses. The marble cladding in this location shows signs of cracking. It is almost impossible to assess accurately the margin of safety against failure of the masonry, but the consequences of failure would be catastrophic. The second problem is the stability of the foundations against overturning.

The approach of the commission to stabilisation of the tower has been a two stage one. The first stage has been to secure an increase in the margin of safety against both modes of failure as quickly as possible by means of temporary measures. Having achieved this, the second stage is to develop permanent solutions recognising that time would be required to carry out the necessary investigations and trials. Significant progress has been made with the first stage. It is a prerequisite of restoration work that temporary works should be non-invasive, reversible and capable of being applied incrementally in a controlled manner.

Temporary stabilisation of the masonry

The masonry problem has been tackled by binding lightly prestressed plastic covered steel tendons around the tower at the first cornice and at intervals up the second storey

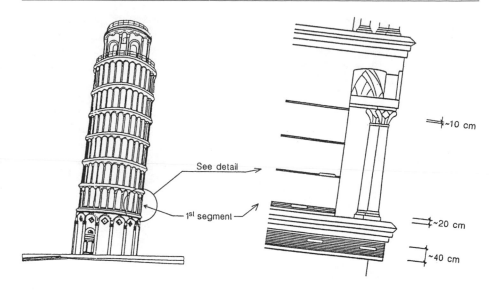

Figure 4.11 *Temporary stabilisation of the masonry with light circumferential prestressing.*

(Fig. 4.11). The work was carried out in the summer of 1992 and was effective in closing some of the cracks and in reducing the risk of a buckling failure of the marble cladding. The visual impact has proved to be negligible.

Temporary stabilisation of the foundations

As mentioned previously, the observation that the northern side of the foundation had been steadily rising for most of this century led to the suggestion that application of load to the foundation masonry on the north side could be beneficial in reducing the overturning moment. Clearly such a solution would not have been considered if it had not been recognised that leaning instability rather than bearing capacity failure was controlling the behaviour of the tower or if the north side of the foundation had been settling.

Before implementing such a solution it was obviously essential that a detailed analysis should be carried out. The purpose of such an analysis was two-fold:
- to ensure that the proposal was safe and did not lead to any undesirable effects and
- to provide a best estimate of the response against which to judge the observed response of the tower as the load was being applied.

A detailed description of the analysis is given by Burland and Potts (1994) who found that a satisfactory result was only forthcoming if the effects of ageing of the underlying Pancone Clay was incorporated in the computer model. The justification for such

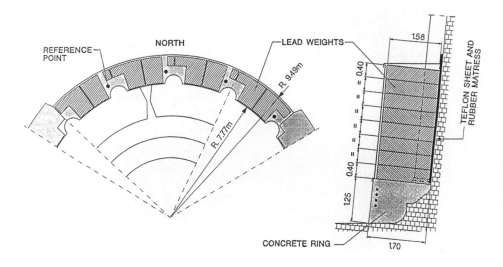

Figure 4.12 *Details of the north counterweight.*

ageing lay in the observed motion of the foundations depicted in Fig. 4.10 as described in the previous section. The computer analysis indicated that it was safe to apply a maximum load of up to 1400 t to the north side of the foundation masonry. Above that load there was a risk that the underlying Pancone Clay would begin to yield resulting in a southward rotation of the tower and excessive settlement of the foundations.

Accordingly a design was developed by Professors Leonhardt and Macchi for the application of a north counterweight and the details of construction are shown in Fig. 4.12. It consists of a temporary prestressed concrete ring cast around the base of the tower at plinth level. This ring acts as a base for supporting specially cast lead ingots which were placed one at a time at suitable time intervals. The movements experienced by the tower are measured with a highly redundant monitoring system consisting of precision inclinometers and levellometers installed on the wall of the ground floor room, high precision levelling of eight survey stations mounted on the wall of the above room, external high precision levelling of 15 bench marks located around the tower plinth and 24 bench marks located along north-south and east-west lines centred on the tower. All the levels are related to a deep datum installed in the Piazza dei Miracoli by the commission.

Observed response

Burland *et al.* (1994) describe the response of the tower to the application of the counterweight. Construction of the concrete ring commenced on 3rd May 1993 and the first lead ingot was placed on 14th July 1993 (Fig. 4.13). The load was applied in four phases with a pause between each phase to give time to observe the response of the

Figure 4.13 *Placing the first lead ingot on 14th July 1993.*

tower. Figure 4.14 shows the construction of the north counterweight. The final phase was split in two either side of the Christmas break. The last ingot was placed on 20th January 1994.

Fig. 4.15 shows the change of inclination of the tower towards the north during the application of the lead ingots as measured by the internal high precision levelling and the inclinometer placed in the north-south plane. The agreement between the two independent monitoring systems is excellent. (Note that Fig. 4.17 does not include the inclination induced by the weight of the concrete ring which amounted to about 4″). It can be seen that the amount of creep between the phases of load is small. However, subsequent to completion of loading, time-dependent northward inclination has continued. On 20th February 1994 (one month after completion of loading) the northward inclination was 33″. By the end of July 1994 it had increased to 4.8″ giving a total of 52″ including the effect of the concrete ring. On 21st February 1994 the average settlement of the tower relative to the surrounding ground was about 2.5mm.

Figure 4.14 *Construction of the north counterweight.*

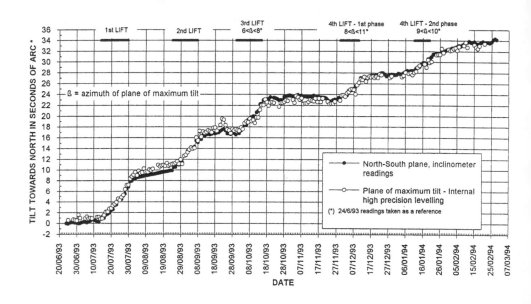

Figure 4.15 *Observed change of inclination of the tower during application of the counterweight.*

Comparison between predictions and observations

Figure 4.16 shows a comparison of the predictions from the computer model and measurements of (a) the changes in inclination and (b) the average settlements of the tower relative to the surrounding ground during the application of the lead ingots. The points in the upper part of Fig 4.16 represent the measured rotations at the end of each phase of loading and the vertical lines extending from them show the amount of creep movement between each phase. For the final phase the creep after one month is shown. It can be seen that the predictions of the computer model give changes in inclination which are about 80% of the measured values. However the predicted settlements are in excellent agreement with the measurements. It is perhaps worth emphasising that the purpose of the computer model was to clarify some of the basic mechanisms of behaviour and it was calibrated against inclinations measured in degrees. The use of the model in studying the effects of the counterweight was to check that undesirable and unexpected responses of the tower did not occur. In this respect the model has proved to be very useful. It has led to a consideration of the effects of ageing and it has drawn attention to the importance of limiting the magnitude of the load so as to avoid yield in

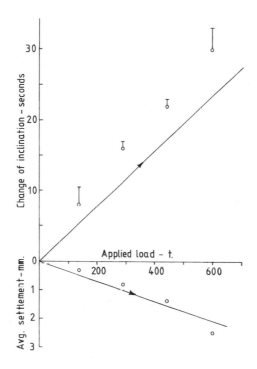

Figure 4.16 *Predicted and observed response of the tower to the application of the counterweight.*

the underlying Pancone Clay. It is perhaps expecting too much of the model for it to make accurate quantitative predictions of movements which are three orders of magnitude less than those against which it was calibrated and the fact that it has done as well as it has is remarkable. The observed movements due to the application of the counterweight have been used to further refine the model.

Permanent stabilisation

For bureaucratic and financial reasons work on the temporary stabilisation of the tower has taken longer than had been hoped. In parallel with these operations the commission has been exploring a variety of approaches for permanently stabilising the tower. The fragility of the masonry, the sensitivity of the underlying clay and the very marginal stability of the foundations has already been referred to. Because of these severe restraints, any measures involving the application of concentrated loads to the masonry or underpinning operations beneath the south side of the foundation have been ruled out. Moreover aesthetic and conservation considerations require that the visible impact of any stabilising measures should be kept to an absolute minimum.

The commission has decided to give priority to so called 'very soft' solutions aimed at reducing the inclination of the tower by up to half a degree by means of induced subsidence beneath the north side of the foundation, without touching the structure of the tower. Such an approach allows the simultaneous reduction of both the foundation instability and the masonry overstressing with a minimum of work on the tower fabric itself.

Some of the key requirements of stabilisation by reducing inclination are as follows:

1. The method must be capable of application incrementally in very small steps.
2. The method should permit the tower to be 'steered'.
3. It must produce a rapid response from the tower so that its effects can be monitored and controlled.
4. Settlement at the south side must not be more than 0.25 of the north side. This restriction is required to minimise damage to the catino and disturbance to the very highly stressed soil beneath the south side.
5. There must be no risk of disturbance to the underlying Pancone Clay which is highly sensitive and upon whose stiffness, due to ageing, the stability of the tower depends.
6. The method should not be critically dependent on assumed detailed ground conditions.
7. The impact of possible archaeological remains beneath the tower must be taken into account.
8. Before the method is implemented it must have been dearly demonstrated by means of calculation, modelling and large scale trials that the probability of success is very high indeed.
9. It must be demonstrated that there is no risk of an adverse response from the tower.

10. Any preliminary works associated with the method must have no risk of impact on the tower.
11. Methods which require costly civil engineering works prior to carrying out the stabilisation work are extremely undesirable for a number of reasons.

After careful consideration of a number of possible approaches the commission chose to study three in detail:

a. The construction of a ground pressing slab to the north of the tower which is coupled to a post-tensioned concrete ring constructed around the periphery of the foundations.
b. Consolidation of the Pancone Clay by means of carefully devised electro-osmosis.
c. The technique of soil extraction as postulated by Terracina (1962) for Pisa and widely used in Mexico City to reduce the differential settlements of a number of buildings due to regional subsidence and earthquake effects. This technique involves the controlled removal of small volumes of soil from the sandy silt formation of layer A beneath the north side of the foundation.

All three approaches have been the subject of intense investigation. Numerical and centrifuge modelling of the north pressing slab have shown that the response of the tower is somewhat uncertain and, if positive, is small while the induced settlements are large. Full scale trials of the electro-osmosis showed that the ground conditions at Pisa are not suited to this method. Both these methods require costly civil engineering works prior to commencement of the stabilisation work. Work on the method of soil extraction is proving much more positive but before describing it a major set-back took place in September 1995 and first this will be described.

A setback

Shortly after the successful application of the temporary counterweight a view emerged that the commission was politically vulnerable and that something needed to be done that would clearly demonstrate the effectiveness of the work so far. There was also considerable concern amongst some members that, should the commission cease to exist, the unsightly lead counterweight would be left in position for many years. Therefore a scheme was developed to replace the lead weights with ten tensioned cables anchored in the lower sands at a depth of about 45m as shown in Fig. 4.17. Additional benefits of this proposal were seen to be that the increased lever-arm would give a slightly larger stabilising moment than the lead counterweight and tensions in the anchors could be adjusted to 'steer' the tower during implementation of induced subsidence. It is important to appreciate that this ten anchor solution was always intended to be temporary.

The major problem with the ten anchor solution is that the anchors have to react against a post-tensioned concrete ring around the tower foundation and this involves excavation beneath the catino at the south side – an operation of the utmost delicacy since it is below the water table. Various schemes for controlling the water were consid-

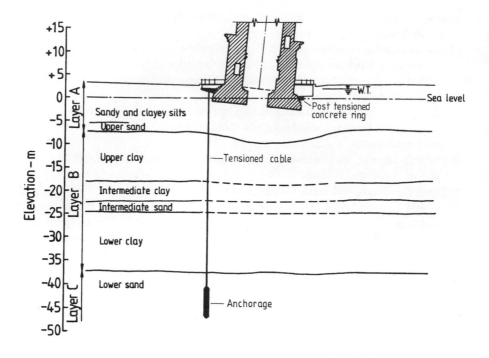

Figure 4.17 *Schematic diagram of the temporary ten anchor solution.*

ered and it was decided to employ local ground freezing immediately beneath the catino floor but well above foundation level. The post-tensioned concrete ring was to be installed in short lengths so as to limit the length of excavation open at any time.

Shortly before commencement of the freezing operation exploratory drilling through the floor of the catino revealed the existence of an 80 cm thick ancient concrete (conglomerate) layer which had evidently been placed by Gherardesca in 1838. There are no archaeological records of this conglomerate and its discovery came as a complete surprise. A key question was whether it was connected to the tower. Exploratory drilling was carried out to investigate the interface between the conglomerate and the masonry foundation. A circumferential gap was found all around the foundation and it was concluded that the conglomerate was not connected to the masonry. Work then started on installing the post-tensioned concrete ring.

Freezing commenced on the north side and the northern sections of the ring were successfully installed. The freezing operations consisted of 36 hours of continuous freezing using liquid nitrogen followed by a maintenance phase when freezing was carried out for one hour per day so as to control the expansion of the ice front. Some worrying southward rotation of the tower did take place during freezing at the north but this was recovered once thawing commenced. Of far greater concern was the discovery of a large number of steel grout-filled pipes connecting the conglomerate to the masonry foundation. These were installed by Girometti in 1934 when the foundation masonry was grouted. In none of the engineering reports of the time is there any reference to these grout pipes or the conglomerate.

In September 1995 freezing commenced on the south- west and south-east sides of the foundation. During the initial 36 hours of continuous freezing no rotation of the tower was observed. However, as soon as the freezing was stopped for the maintenance phase the tower began to rotate southward at about 4 arc seconds per day. The operation was suspended and the southward rotation was controlled by the application of further lead weights on the north side. The resulting southward rotation of the tower was small, being about 7 arc seconds, but the counterweight had to be increased to about 900 t. The main concern was the uncertainty about the strength of the structural connection between the conglomerate and the masonry formed by the steel grout pipes. In view of this uncertainty the freezing operation was abandoned and work on developing the permanent solution was accelerated.

Induced subsidence by soil extraction

Figure 4.18 shows the proposed scheme whereby small quantities of soil are extracted from layer A below the north side of the tower foundation by means of an inclined drill. The principle of the method is to extract a small volume of soil at a desired location leaving a cavity. The cavity gently closes due to the overburden pressure causing a small surface subsidence. The process is repeated at various chosen locations and very gradually the inclination of the tower is reduced.

Two key questions had to be addressed:
- Given that the tower is on the point of leaning instability, is there a risk that extraction of small quantities of soil from beneath the north side will cause an *increase* in inclination?
- Is the extraction of small volumes of ground in a controlled manner feasible, will the cavities close and what is the response at the soil/foundation interface?

The first issue has been studied in great detail using two independent approaches – numerical modelling and physical modelling on the centrifuge. The numerical model described previously was used to simulate the extraction of soil from beneath the north side of the foundation. Even though the tower was on the point of falling over it was found that, provided extraction takes place north of a critical line, the response is always positive. Moreover, the changes in contact stress beneath the foundations were small. Advanced physical modelling was carried out on a centrifuge at ISMES in Bergamo. As for the numerical modelling, the ground conditions were carefully reproduced and the model was calibrated to give a reasonably accurate history of inclination. The test results showed that soil extraction always gave a positive response.

The results of the modelling work were sufficiently encouraging to undertake a large scale development trial of the drilling equipment. For this purpose a 7 m diameter eccentrically loaded instrumented footing was constructed in the Piazza north of the baptistry (Fig. 4.19). The objectives of the trial were:
1. To develop a suitable method of forming a cavity without disturbing the sur-

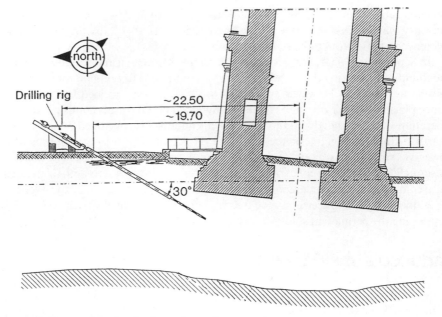

Figure 4.18 *Induced subsidence by soil extraction.*

rounding ground during drilling.

2. To study the time involved in cavity closure.
3. To measure the changes in contact stresses and pore water pressures beneath the trial footing.
4. To evaluate the effectiveness of the method in changing the indination of the trial footing.
5. To explore methods of 'steering' the trial footing by adjusting the drilling sequence.
6. To study the time effects between and after the operations.

It must be emphasised that the trial footing was not intended to represent a scale model of the tower.

Drilling is carried out using a hollow-stemmed continuous flight auger inside a contra-rotating casing. When the drill is withdrawn to form the cavity an instrumented probe located in the hollow stem is left in place to monitor its closure. A cavity formed in the layer A material has been found to close smoothly and rapidly. The stress changes beneath the foundation were found to be small. The trial footing was successfully rotated by about 0.25° and directional control was maintained even though the ground conditions were somewhat non-uniform. Rotational response to soil extraction was rapid taking a few hours. Very importantly, an effective system of communication, decision taking and implementation was developed.

The soil extraction trial was successfully completed in March 1996 and it was argued that there was nothing more that could reasonably be done to prove the effective-

Figure 4.19 *Soil extraction trial showing 7 m diameter eccentrically-loaded footing and inclined drill.*

ness of soil extraction short of testing it on the tower itself. After much debate the decision was taken by the commission to carry out preliminary soil extraction beneath the north side of the tower with the objective of observing the response of the tower to a limited and localised intervention. A safeguard structure was to be constructed in the form of a horizontal cable stay attached to the tower at the third storey which could be tensioned to steady the tower in the event of detrimental movements.

Soil extraction from beneath the tower

In August 1996 the decree giving the commission its mandate fell and was not renewed so that further work ceased. In December 1996 the commission was disbanded and a new one (the seventeenth) was established with a substantial change of membership. The new commission first met in June 1997 and it was decided that a complete review of all possible stabilisation measures should be undertaken. The case for soil extraction had to be argued again from scratch. In July 1998, after a year of heated debate, the decision was again taken to implement preliminary soil extraction to assess its effectiveness in reducing the inclination of the tower.

By December 1998 the temporary safeguard cable stays were in place. Preliminary soil extraction was to be carried out over a limited width of 6 m using twelve boreholes. A target of a minimum of 20 arc seconds reduction in inclination was set as being large enough to demonstrate unequivocally the effectiveness of the method.

Because of the extreme delicacy of the project only 20 litres of soil were to be extracted every two days with continuous real time monitoring to assess the results. A strict system of command and control was established in which each extraction operation required an instruction based on the response of the tower and signed by the responsible officer.

On 9th February 1999, in an atmosphere of great tension, the first soil extraction took place. For the first week the tower showed no discernible response but during the following days it began very gradually to rotate northwards. As confidence grew the rate of soil extraction was increased. At the beginning of June 1999, when the operation ceased, the northward rotation was 90 arc seconds and by mid-September it had increased to 130 arc seconds which is equivalent to about 40 mm at the top. At that time three of the 97 lead ingots (weighing about 10 t each) were removed and the tower exhibited negligible further movement.

During preliminary under excavation, soil extraction mainly took place outside the footprint of the foundation and locally only extended beneath the north edge of the foundation by about 1.5 m. It is of interest that the southern edge of the foundation was observed to rise by about one tenth of the settlement at the north. This may be contrasted with the numerical model which predicted small settlements at the south. The reason for this difference may be due to the fact that a plane strain model was used whereas the soil extraction process is highly three dimensional.Whatever the reason, the uplift at the south is highly beneficial as the volume of soil to be extracted is reduced and it seems likely that reduction of stress is taking place in this critical region.

Having demonstrated that soil extraction produced a positive resposse, the commission formally approved the application of the method for permanent stabilisation. Using 41 extraction tubes, work on the full intervention commenced on 21st February 2000. It is estimated that it will take about eighteen months of careful soil extraction to reduce the inclination of the tower by about half a degree which will be barely visible. At the time of writing (November 2000) a reduction of inclination of 1350 arc seconds has been achieved and the pattern of uplift at the southern edge has been maintained.

Concluding remarks

The stabilisation of the tower of Pisa is a very difficult challenge for geotechnical engineering. The tower is founded on weak, highly compressible soils and its inclination has been increasing inexorably over the years to the point at which it is about to reach leaning instability. Any disturbance to the ground beneath the south side of the foundation is very dangerous. Therefore the use of conventional geotechnical processes at the south side, such as underpinning, grouting, etc., involves unacceptable risk. The internationally accepted conventions for the conservation and preservation of valuable historic buildings, of which the Pisa tower is one of the best known and most treasured, require that their essential character should be preserved, with their history, craftsmanship and enigmas. Thus any invasive interventions on the tower have to be kept to an absolute minimum and permanent stabilisation schemes involving propping or visible support are unacceptable and in any case could trigger the collapse of the fragile masonry.

The technique of soil extraction provides an ultra-soft method of increasing the stability of the tower which is completely consistent with the requirements of architectural conservation. Different physical and numerical models have been employed to predict the effects of soil removal on the stability.

The soil extraction intervention, only undertaken after having been satisfied by comprehensive numerical and physical modelling together with a large scale trial, has demonstrated that the tower responds very positively. There is still a tense journey ahead for the tower, requiring detailed communication and control and the utmost vigilance, but indeed the first step has been taken in the permanent geotechnical stabilisation.

References

Burland, J.B. 1990. *Pisa Tower. A simple temporary scheme to increase the stability of the foundations*. Unpublished report to the Commission Q.

Burland, J.B. and Potts, D.M. 1994. Development and application of a numerical model for the Leaning Tower of Pisa. *Int. Symp. on Pre-failure Deformation Characteristics of Geo-materials*. 1S-Hokkaido 1994, Japan, 2, pp. 715–38.

Burland, J.B. and Viggiani, C. 1994. Osservazioni sul comportamento della Tone di Pisa. *Rivista Italiana di Geotecnica*, 28, 3, pp. 179–200.

Burland, J.B., Jamiolkowski, M., Lancellotta, R., Leonards, G.A. and Viggiani, C. 1994. Pisa update – behaviour during counterweight app cation. *ISSMFE News*, **21**, No 2.

Hambly, E.C. 1985. Soil buckling and the leaning instability of tall structures. *The Structural Engineer*, **63A**, 3, pp.77–85.

Ministero dei Lavori Pubblici. 1971. *Ricerche e studi sulla Torre di Pisa ed i fenomeni connessi alle condizione di ambiente.* 3 vol., I.G.M., Florence.

Potts, D.M. and Gens, A., 1984. The effect of the plastic potential in boundary value problems involving plane strain deformation. *International Journal for Numerical and Analytical Methods in Geomechanics*, **8**, pp. 259–86.

Schofield, A.N. and Wroth, C.P., 1968. *Critical State Soil Mechanics*. McGraw-Hill Book Co., London.

Terracina, F., 1962. Foundations of the Tower of Pisa. *Geotechnique*, **12**, 4, pp. 336–9.

Postscript

At the time of going to press, the intended reduction of inclination of 0.5° has been achieved which means that the tope of the tower has moved northwards by 50 cm – not enough to be visible. The lead counterweight has been removed as have the temporary safeguard cables. To improve stability, the ancient concrete ring in the floor of the catino has been connected to the foundations of the tower. On 16th June 2001 the tower was handed back to the civic authorities and it is anticipated that it will be opened to the public during 2001. It is intended that the movements of the tower will continue to be closely monitored so as to establish its new pattern of seasonal behaviour. Only time will tell whether the previous progressive seasonal movements have been eliminated. Even if they have not, it is likely to be hundreds of years before the tower returns to its 1990 inclination and if necessary, the process of soil extraction could be repeated.

5 Applying lessons from failures to management and design

Jonathan G.M.Wood

Introduction

Over the last 35 years I have worked on the investigation of 'failures' of silos, box girders, bridges, tunnels and buildings and a miscellaneous assortment of other structures and objects. For the last 15 years much of this work has been on structures where the deterioration has been a major contributory factor. The details of these investigations were in reports to clients, some of which were used in litigation. Some clients and sometimes the courts require total confidentiality. Other clients have given permission for and on occasions encouraged, the use of results directly or indirectly in publications or to assist in the development of standards and industry guidance, some of which are referenced below.

'Failures' range in severity, being difficulties, problems, failures or collapses. It is worth distinguishing between the general run of difficulties and problems which are investigated and resolved in the normal course of engineering practice and the more limited range of failures and collapses which attract the attention of the media and lawyers. If we can identify and analyse the difficulties and problems, particularly where there were 'near misses', and rapidly communicate the lessons to those dealing with similar situations, we can substantially reduce the risks of major failures and collapses, as well as the hazards of litigation. This will also help the process of successful innovation.

Lines of communication

The routes and timescale for transferring information from failures to working engineers are uncertain. When the rare major 'political' structural collapses occur they have often triggered substantial programmes of investigation, research and accelerated rewriting of standards and codes (e.g. Ronan Point and Milford Haven box girder collapse). Lesser failures are soon forgotten and the lessons are often not learned. Improved communication of data from failures to practising engineers, those who write codes and guidance and the research and teaching community is needed.

Lessons from failures must be communicated both within professions and between specialisations. This becomes more difficult as new sub-specialisations are regularly spawned and develop their own distinct languages. The unclear contractual interfaces and communication difficulties between specialisations are in themselves a frequent contributory factor to failures. Similarly there is a need to communicate across the generations highlighted by a recent graduate's remark 'What was Ronan Point?'

Structural failures

Most structural failures occur deliberately in laboratory testing. The enormous body of published research data provides an essential yardstick of the state of the art for forensic investigation. There is a well-established, but slow, route from structural research via publication in research or professional institution journals into the drafting of standards and codes of practice and industry guidance from institutions such as CIRIA, BRE, TRL, Steel Construction Institute, Concrete Society and the wide range of other specialist organisations, as well as commercial documentation on specific products. Eurocodes have extended this chain of communication. This body of received good practice is the reference source for designers and contractors and their suppliers in their endeavour to act reasonably and minimise liabilities.

In the best engineering teams this codified and documented good practice is supplemented by the caution and experience of older members of the team in design or review and from the resident engineers or contractor's staff considering the practicalities and safety of construction. However too frequently the letter of the codes, now often summarised in a computer package, is followed without considering if the structure and application are within the range considered by those who drafted the documents. This computerised code compliance works reasonably well for conventional structures, in part because of the generosity of the factors of safety used. However, when unusual conditions or structures are involved problems can occur. When they do we must create a fast track to upgrade the codes and guidance.

Durability failures

The poor durability of many 20th century innovations in materials have become a matter of increasing concern and expense to owners (Wood, 1996; Gerwick, 1994). Durability research is inherently more difficult because the processes are long term, with complex interactions between materials and the microclimate, so they cannot be accelerated without distortion. Much short term durability research has misled. One can make and break a steel structure in a few days and a concrete one in a month, but durability tests take decades so we have to look for structures that have matured for decades and analyse their problems, preferably before they collapse (Woodward and Williams, 1988). The information necessary for improving design must largely come from the rigorous analysis of problems and failures in the field, with complementary laboratory work (Wood and Crerar, 1995; Wood et al., 1996). Much more work is needed in analysing durability failures and applying the lessons to design and specification if we are to avoid the waste of resources and environmental cost of premature reconstruction of our 20th century infrastructure.

Organisational failures

In all investigations there is the narrow technical question of identifying the physical, chemical and occasionally biological reasons for the failure. There are also wider questions of why the circumstances which lead to the technical fault arose (Walker, 1980; Royal Commission, 1971). These often relate to the gaps or ambiguities in the allocation of responsibilities for specification, design, construction, operation and maintenance or the inadequacies of time and resources for the work. Limitations in the technical competence of those involved and the availability of information to them are also recurrent features of failures.

Impediments to proper communication

There is a natural human instinct to avoid association with failures and to seek to minimise their importance and impact. Even within major construction organisations, instances of failure are often hushed up, rather than being used to highlight the organisational and technical causes, so that all learn from the unpleasant experience and are less likely to repeat it.

Commercial organisations have a similar approach and a tendency to divert attention from problems with products they are associated with, but they emphasise those of their competitors. These instincts, to preserve market share and to communicate only the positive, spill over into and distort the deliberations of the industry committees on standards and guidance documents. Most people now apply a degree of scepticism when hearing of tobacco industry research on cancer and similar scepticism is prudent in evaluating construction research. I have drawn attention to these realities not because I think one can alter human nature or commercial instinct, but to highlight the need for data from rigorous analysis to balance the commercial views. This is essential if we are to meet the long-term interests of the customers of the construction industry. These customers, including governments, must provide more funding to facilitate this.

The customers of the construction industry, and many organisations in the construction industry, tend to react hastily to particular problems with structures and materials. Frequently there is a general ban or an aversion to the use of designs or materials with any indirect link to the early anecdotal reports of a failure. There is a heavy cost to society from these disproportionate and unselective reactions which lead to the abandonment of established technologies in favour of the new and untried, which after a few years reveal their own problems. Promptly reported accurate diagnosis of the faulty details of design, materials and/or construction, and methods of avoiding the recurrence of problems by proper development will save resources and reduce risk.

Forensic aspects

The relatively narrow term forensic engineering means literally, used in courts of law. However, most investigative work is carried out to diagnose and to identify solutions. Once investigation becomes overlaid by the legal process, which is adversarial and focuses on blame attribution, it becomes more difficult to draw balanced conclusions on the technical and organisational problems that led to failure. Litigation can also delay or limit the dissemination of information. In some instances technical conclusions drawn by legal tribunals are wrong and this becomes apparent in reviews of failures using data from later research (Burland, 2000). There could be considerable benefit from reviews of past failures with technical hindsight. This might be more beneficial than some of the 'foresight' exercises that have recently been diverting government.

Specific testing related to the features of a structure with problems can make a substantial contribution to available knowledge. Sometimes it needs the authority of a QC in litigation to justify the expenditure on rigorous testing (e.g., Wood *et al.*, 1988). In litigation good test data is always stronger than theory and opinion, but if it is only available to the litigants the opportunity to reduce risk in the wider construction community is lost.

There is a risk that forensic engineering will become yet another distinct specialisation (Carper, 1998), with its own publications and conferences, apart from the wider established forums of the construction industry. Basic investigative skills should be one of the normal capabilities of all chartered engineers and training in them and in the lessons from failures should form part of all undergraduate courses and continuing professional development. Each branch of engineering needs to have practising members who have further developed their investigative skills, often by working with the research community, in their specialities. Those who have investigated problems and failures can best assess and communicate the lessons in evolving improved designs. They need to ensure that this information is also packaged and routed into the development of standards and into the teaching process.

Innovation, failure, ostracism and rediscovery in construction

The changing fashions in the construction industry show cycles of enthusiastic innovation, then problems and failures, leading to abandonment and then with a new generation, rediscovery. Many clients are risk-averse and misguidedly avoid innovation. This aversion to innovation is not good for the industry or for clients, as proper innovation will reduce risks and improve value.

The development and popularisation of a new structural form or material is often based more on enthusiasm than research. Then inadequate analysis and reporting of the difficulties and problems inevitably associated with innovation lead to misunderstandings and a general ban, prejudice or a reluctance to use the form. The problems attract the attention of the universities leading to programmes of research into the material and its use. This knowledge explosion often coincides with the commercial death of the structural form. Examples of this include box girder bridges (ICE, 1972; Rockey and Evans, 1981), post-tensioned concrete bridges, high aluminum cement, epoxy coated reinforcement, GRC. Then 20 or 30 years later the pendulum swings back. The true potential of the material, when properly used, then becomes apparent and it enjoys a rejuvenation in the market place. If we are to avoid the economic damage that these wide swings in construction fashion produce, we need more rigorous step by step development of innovative structures and materials so that they are not used inappropriately. The early warning signs of their limitations need to be identified, investigated and communicated.

One result of innovation problems has been the growth of a large list of 'excluded materials' in construction contracts. The UK Building Research Establishment and some property owners are now seeking to review and correct these lists (BRE, 1997; BCO/BPF, 1997). Many of the problems related to the misuse of materials in specific adverse conditions, but the materials can be used satisfactorily in many applications. With our improving methods of analysing failures, particularly those associated with materials, we need to review old reports and revise and refine earlier conclusions. Even where original investigations correctly identified causes, the public perception often retains the misunderstandings from early reports (e.g. misattribution of early cases of alkali-aggregate reaction to marine source of aggregate, 'HAC' failures which were largely due to poor details and tolerances). To obtain a balanced picture one must combine 'failure' investigations with 'success' investigations of proven use.

Broadening the objectives and budgets of failure investigation

The primary questions in investigation are 'What has gone wrong?', 'How do we put it right?' and then 'Who pays for it?', at which stage it tends to become forensic with lawyers involved! These are generally regarded as closed questions within the frame-work of a specific contract and often there is no budget, time or inclination from the parties that the wider lessons should be learnt.

For the more senior members in contracting and consulting engineering organisa-tions there should be a commercial concern to ensure that the lessons from any prob-lem are properly communicated within their organisations and the industry, as the financial penalties and, where safety is concerned, the legal penalties, of a further problem may be more major. They should ensure that this is resourced and funded. For engineers who are members of the professional institutions there is also a need to communicate information on particular hazards to their fellow professionals through SCOSS (Standing Committee on Structural Safety) or by the publication of notes or reports in journals.

Where public safety or the safety of employees is concerned the heavier duties imposed by health and safety legislation arise. There is a conflict between the obligations of companies and individuals to act reasonably to properly communicate risks to their employees and the public, and their wish to keep commercially confidential the risks associated with their products. When parties to litigation settle and agree to keep all documents and test data confidential they may be failing in their duty to act reasonably to minimise risks to their employees and the public.

Deficiencies in standards and guidance

The law has two yardsticks for engineering performance – reasonable good practice and fitness for purpose.

Many failures I have investigated have been due to ignorance of reasonable good practice by those involved in design and/or construction. Sometimes this is because the contractual arrangements have given responsibilities to those with inadequate training, experience or qualifications. This is often a consequence of design and build contracts taken on by those with experience of building, but not design.

A more difficult situation arises for the working engineer and in litigation, when structures fail to be 'fit for purpose' despite complying with 'reasonable good practice' as set out in British Standards, BRE Digests, CIRIA Reports and documents produced by commercial bodies. This can occur when revisions to standards lag behind developing knowledge and innovations in design. It also occurs when standards make simplifications appropriate to a limited range of conditions without making this clear. Accelerating the feedback from construction problems to the codes and guidance documents to eliminate these flaws in codes should have high priority.

Loading

Developments in high-rise racking and rolling library storage have increased loadings on structures and their sensitivity to deflections. Because information in BRE Information Paper 19/87, which recommended a revision of BS 6399:1984, was not consulted by a specifier who relied just on the BS, a warehouse floor had to be rebuilt at the cost of over £1,000,000.

The limitations of BS153 HA loading for long span bridges, which became apparent during the Merrison checks in the 1970s on the Severn Bridge, showed that it underestimated traffic loading by a factor of three, even before the effects of increased permitted vehicle weights. When the load intensity was estimated in the 1940s it was assumed that vehicles would be spaced further apart on long spans. Should the designers in the 1960s have checked this when embarking on a project of such magnitude? In that instance the Department of Transport as client was able to rapidly bring in the revised loadings for appraisal and for design to BS5400. This rapid reaction will become more difficult as the Highways Agency seeks to shed responsibility for design standards by limiting contract requirements to broad performance objectives (Pickett, 1998).

Old structures

Appraisals of older structures frequently reveal details, original design assumptions or structural problems requiring investigation. Sometimes changes in use or our better knowledge of structural behaviour show old structures to have a lesser factor of safety than originally intended and that now required. The IStructE *Appraisal of Existing Structures* (1996) provides an excellent summary of information gained from such investigations.

Durability

Standards relating to the durability of 20th century materials, like reinforced concrete, have a particularly poor record of delivering fitness for purpose. There have been improvements as codes have developed, in this case CP114 developed via CP110 to BS8110 and now Eurocodes. However the lack of a factor of safety in durability design, the industry pressure to simplify specification and minimise cost and the poor detailing of designs which create aggressive conditions and make the achievement of correct covers and compaction difficult, will ensure that durability problems will continue to plague building owners.

The detailed analysis of the slow realisation of the industry of the problems from chlorides in concrete and progressive changes in standards in UK and US has been published by CIRIA (Pullar-Strecker, 1987). This well illustrates the slow timescale from the realisation of problems to the application of data from failure investigation in standards. For concrete in the most severe conditions of chloride exposure these developments continue (Wood,1997).

Alkali aggregate was first identified as causing serious structural problems in the UK at Charles Cross Car Park in Plymouth in 1975. Data from detailed investigations of the many structures which have been damaged by AAR, some of which led to litigation, have substantially contributed to the development of recommendations for dealing with structures with AAR (I Struct E, 1992; Wood and Johnson, 1993) and in the developing the recommendations for minimising the risk in new construction (Hawkins, 1995). However the current specification recommendations for general construction need specific development for major projects (Wood, 1993) and long life structures particularly for the severe conditions in tunnels and dams revealed by investigations.

Silos

Silos are produced from kits of standard parts by design and build contractors for a competitive market which demands ever larger sizes and where price is more important than safety. They are particularly prone to failures (Wood, 1997) and litigation. Failure investigations (see case study) have provided much of the information for improved standards (BS5061, 1974) and guidance (BSI, 1987).

Format for communication of failures

SCOSS has done excellent service in reviewing and publicising a wide range of structural risks and its periodic summary reports (SCOSS, 1997) are widely circulated and read. The professional magazine *New Civil Engineer* (NCE) provides early information on major failures worldwide, but the detail that emerges from full investigation is seldom available until much later, if at all. NCE (Parker, 1998) are advocating better reporting of failures.

To better communicate the information from failures throughout the construction industry we must bridge the gap between those with the knowledge of failures to the specific individuals who need to know now or in the future. This may be done directly, as a summary, or indirectly by the updating of the standards and guidance on which most engineers rely.

We are all swamped with information already and are only interested in the failures which relate to the work with which we are currently involved. Unless we use information technology to sort out the subsets of failure details which relate to the work of particular groups of engineers we will not successfully deliver the information. Similarly information technology will enable us to deliver these selected sets more quickly.

The essential requirement is the distillation of the key facts from the mass of information which start to accumulate when things go wrong. For most of the investigations I have been concerned with this could be achieved as follows:

Summary, keyword style, three lines maximum referring to
Data sheet, one sheet of paper, text with up to three photos or diagrams referring to:
Published reports, (e.g.Chapman, 1998) based on and possibly referring to the
    existence of:
Detailed reports, possibly confidential, prepared on the failure.

Selected sets of data sheets could be bound into design and site office sets related to particular activities, especially where hazards are known to be high (e.g. crane failures during site operations, erection stages of major bridges, bored tunnelling operations, NATM tunnelling, punching failures of flat slab structures, corrosion in prestressing, weld failures, trench wall collapses, barrier failures on car parks and bridges, etc.). The most important historical cases and international experience should be included.

SCOSS would be a suitable organisation to set the format and ground rules for the summary and data sheets and their dissemination in traditional and/or electronic format. A possible format, with data on the first collapse I investigated, is appended to this chapter as a case study. Using this format, many organisations could prepare data sheets, and subject to review for accuracy, make them available.

Difficulties would arise where there were inadequate data to clarify the causes of failure and/or a dispute. This could be overcome if the ground rules for the data sheet required agreed facts to be separated from a list of all possible causes of the failure, with reference to any documents supporting particular views.

Soon after a significant incident a provisional data sheet could be produced which could be revised as each stage of the investigations were concluded. This would accelerate the dissemination on items now covered by SCOSS's periodic summaries.

Conclusions

There are substantial benefits to the construction industry and to its customers from improving the quality and speed of communication of information from failures, whether they are difficulties, problems, failures or collapses.

The analysis of the causes of failures must include the organisational and contractual circumstances and staff training and qualification which lead to technical failures, as well as the technical analysis.

Successful innovation and the orderly development of contracts and site organisations, designs and construction has been disrupted and distorted by

- lack of properly funded development stages of innovation, with proper diagnosis of early difficulties and problems before major failures occur.
- failures to identify and find remedies for the detailed localised causes of failures leading to a generalised prejudice and contractual exclusion against sound engineering designs and materials.
- the slow process of developing quality guidance and standards, and rapidly and correctly incorporating the lessons of failures in them.

The development of a standard format for a failure data sheet and procedures for electronic sorting and dissemination, could provide a means of reducing risks and speeding the development of improved contracts, designs, materials and construction procedures.

The development of an accurate database of important failures worldwide would be an invaluable resource in university teaching, continuing professional development, training in design offices and on site, and for the organisations which develop design guidance and standards.

Case study

Failure data sheet: farm forage silo collapse, 1965

Summary: A bolted vitreous enamel steel sheet forage tower silo 15 m high, 5.6 m diameter, collapsed following the progressive shearing of bolts up the vertical lap bolted joint. Silage pressure in excess of design values and stress concentrations on bolts led to failure initiation.

Structure: The silo was one of a pair constructed on a dairy farm in the Scottish borders in 1963. The cylindrical silos 5.6 m diameter and 15 m high were constructed

Figure 5.1 *Collapse following zip failure of vertical bolted seam.*

Figure 5.2 *Location of initial bolt failures.*

of vitreous enamelled steel sheets (1.8 m × 0.9 m) with vertical and horizontal bolted lap joints. The sheet thickness was 4.7mm for lower rings to 2.4 mm for upper section. One sheet per ring had a large manhole and door which was immediately adjacent to the bolted joint at lower levels.

Use: The two silos had been successfully used in 1963 and 1964 for young, slightly wilted grass but with some effluent. Both silos had been filled in 1965 and after settlement further filling was planned. The silo which failed had been filled in May and June with three layers of young grass, the lowest of which was 75 to 80% moisture content, at higher levels it was drier (60 to 70% moisture content). There was a large amount of effluent seepage exuded from joints at the lower levels.

Failure sequence: At 07.30 and again at 11.30 on 2nd August two bolts at the double lap between vertical and horizontal joints between rings two and three up had failed in shear. The progressive spread of bolt shear failures down and up to 6m was observed between 13.50 and 14.20 when the silo collapsed (*see* Fig. 5.1). Figure 5.2 shows, on the other silo, the location of the bolts which initially failed.

Assessment: The silo was designed for fibrous maize and alfalfa crops which can be wilted to part dry in US conditions. The young compressible grass and more difficult weather conditions in UK gave much higher pressures. The bolted joint strength, designed assuming uniform load per bolt, was reduced by the drilling of holes for a cylindrical geometry but assembly as conical, the manhole adjacent to joint, the use of fully threaded bolts which sheared on the weaker threaded section, some corrosion from acid effluent. (Report by W.F. Cassie and J.G.M. Wood 12.1.66, details from SS&D, Northbridge House, Chiddingfold GU8 4UU; *see* Figs. 5.1 and 5.2)

Actions: Report on failure to manufacturer used to improve designs and instructions for use. Data from collapse was used in research (PhD Thesis, J.G.M. Wood, Univ. of Newcastle-on-Tyne 1970) and in developing BS5061:1974 for forage tower silos. See 'Greater Safety on the Farm – a standard for tower silos', *BSI News*, May 1974.

References

BCO/BPF 1997. Good practice in the selection of construction materials. BCO/BPF.

BRE. 1997. List of excluded materials : a change in practice, *BRE Digest 425* , BRE, London.

BS5061 1974. *Specification for Cylindrical Forage Tower Silos and Recommendations for their Use.* BSI London.

BSI/BMHB 1987. *Silos: Draft Design Code* BSI and BMHB, London.

Burland, J. 2000. Ground structure interaction: Does the answer lie in the soil?, *The Structural Engineer*, **78**, pp 23–24.

Carper, K.L. 1998. Current structural safety topics in North America, *The Structural Engineer*, **76**, 12.

Chapman, J.C. 1998. Collapse of the Ramsgate Walkway. *The Structural Engineer,* **76**, 1.

Gerwick, B.C. 1994. The Economic Aspects of Durability – How much added expense can be justified? *Proc. Symp. Durability of Concrete* (ed. Khayat, K.H.) pp. 1–19, ACI/ CANMET, Nice.

Hawkins, M.R. (Chairman), 1995. *ASR – Minimising the Risk of Damage to Concrete.* Guidance Notes and Model Specification Clauses, Concrete Society TR30.

ICE. 1972. Steel Box Girder Bridges, *Proc. Int. Conf.* ICE, London.

I Struct E 1992. *Structural Effects of Alkali-Silica Reaction,* Technical Guidance on Appraisal of Existing Structures, SETO, London.

I Struct E 1996. *Appraisal of Existing Structures,* 2nd edition, SETO, London.

Parker, D. 1998. Speaking out for Safety. *New Civil Engineer,* 9/16 , pp 20–21.

Pickett, A. 1998. Global Overview. *IABSE Henderson Colloquium on Design.*

Pullar-Strecker, P. 1987. Appendix to *Corrosion Damaged Concrete, Assessment and Repair,* CIRIA Butterworths.

Rockey, K.C. and Evans, H.R. 1981. *The Design of Steel Bridges,* Granada, London.

Royal Commission. 1971. Report of the Royal Commission, *Failure of West Gate Bridge,* Melbourne.

SCOSS 1997. Report Summary *Structural Safety 1994–96.*

Walker, A.C. 1980. Study and Analysis of 120 Cases of Structural Failures. *Proc. of Symposium on Structural Failures in Buildings,* IStructE, London.

Wood, J.G.M. 1993. Some Overseas Experience of Alkali Aggregate Reaction and its Prevention: Specification for Major Projects: Bridges, Tunnels and Dams. *ITBTP Annales.* Paris, No 518.

Wood, J.G.M. 1996. Durability design: applying data from materials research and deteriorated structures, in *Bridge Management 3* (ed. Harding, J. *et al.*), Spon.

Wood, J.G.M. 1997. Silos: Evolution by Failure. *Structural Engineering Int.*

Wood, J.G.M. 1997. Achieving Durable Concrete, in Prediction of Concrete Durability, *Proc. STATS 21 st Conf.* London, Nov. 1995 (eds. Glanville, J. and Neville, A.), Spon.

Wood, J.G.M. and Johnson, R.A. 1993. The Appraisal and Maintenance of Structures with Alkali Silica Reaction. *The Structural Engineer,* **71**, 2, pp. 19–23.

Wood, J.G.M. and Crerar, J. 1995. Analysis of chloride ingress variability for the Tay Road Bridge, *Structural Faults & Repair – 95.* Forde, M.C. (Ed.) Edinburgh.

Wood, J.G.M., Nixon, P. and Livesey, P. 1996. Relating ASR damage to concrete composition and environment, *Alkali-Aggregate Reaction in Concrete* (10th Int. Conf. Melbourne, ed. Shayan, A.) pp. 450–7.

Wood, J.G.M., Chrisp, T.M. and Blackler, M. J. 1988. Comparison of Model and Full-Scale Test Results with Simplified and Finite Element Analysis of Eccentrically Discharged Silos. ISO/FIP. In *Int. Conf. on Silos,* Karlsruhe, W. Germany.

Woodward, R.J. and Williams, F.W. 1988. Collapse of Ynys-Y-Gwas Bridge, West Glamorgan. *Proc. Inst. Civ. Engrs,* **1**, 4, pp. 635–69.

6 Learning from failures

J.C.Chapman

Introduction

We learn best by personal experience, but because natural selection is inimical to civilisation, humanity can only progress by vicarious experience and by reason. Whether progress will lead to destruction is the greatest issue of our time, but is not the subject of this book.

The prevention of failures is a modest but worthy aim, and can make a signficant contribution to sustainability as well as to safety. This article, which draws heavily on Chapman (2000*a*), aims principally to describe some failures that the writer has investigated, and to suggest steps towards prevention. However, some more general but related observations and opinions will not be suppressed. Where preventive steps have already been taken, these will be reported with especial pleasure.

The reporting of successes, failures, and near misses, is essential to progress; it is part of an engineer's professional obligation. The significance of minor failures or incidents depends partly on their frequency of occurrence; if they are not reported, the frequency will not be known.

The investigation of failures should be approached with humility and recognition that we all make mistakes. The investigator has the incomparable benefit of hindsight.

Risk assessment

Public perceptions and acceptance of risk vary widely accordingly to context. In general, risk is more acceptable when the activity is volitional or inherently dangerous, as when travelling by road. In some cases warnings are issued to protect the public, as in the Highway Code. In other cases the warning is to protect the provider of the risk – 'share prices may go down as well as up' (in small print), or 'this operation has a 2% mortality rate'.

The purpose of risk assessment for structures is to identify possible forces or events beyond normal design assumptions, and where possible to mitigate their effects. Each contingency, and the design response, is recorded. A key feature of risk assessment is that the designer is obliged to think the unthinkable, as distinct from designing to resist specified loadings; the mental approach is different. Effects to be considered may include, for example, lack of maintenance, faulty operation, strike action, sabotage, extreme weather, collision, faulty equipment, power failure, design error.

Provision may also be made against unidentified risks. For example, provision against progressive collapse requires the structure to survive a degree of damage, but the location or cause of the damage does not need to be specified. By providing connections which are as strong as the parts joined, a degree of robustness against unforeseen events will be provided.

Risk analysis (Tietz, 1988) requires data, notably on winds, waves, temperature, earthquakes, service loading, reliability of equipment, and perhaps human performance. The probability of maximum adverse conditions coinciding can also be estimated. In some cases the result of risk analyses will be embedded in a design code. The design assumptions, incorporating considerations of risk, should be made known to the owner and operator of the structure. In a hierarchy of public risk acceptance, it can be expected that unprovoked collapse of fixed structures on land for public use, will in normal parlance, be regarded as unacceptable.

Ship's derrick

The boom of the derrick was attached to the mast by a hinge and swivel. Vertical movement was allowed by a hinge consisting of a pin passing through three lugs, one of which was between the other two and attached to the boom. The outer lugs were supported on a horizontal plate which could swivel about a vertical pin passing through a horizontal platform attached to the mast. Lubrication of all moving surfaces was obviously vital, especially at the swivel, but apparently the crew found it more convenient to heave on the end of the boom to free the rusted surfaces.

Eventually, the centre lug failed and the falling boom struck the head of a Polish seaman, who suffered permanent incapacitating brain damage. After six years the seaman was awarded substantial damages, through the sustained efforts of a Polish resident.

This was my first experience of the legal process, and its capacity for delay. It was the only collapse to come my way which was caused by neglect of maintenance. It prompts the question of the extent to which designers should make allowance for deficient maintenance, which can happen even when maintenance requirements are clearly specified. If the possible consequences include loss of life, it seems that some allowance should be made in design.

- Operational error or abuse should be considered in the risk assessment.
- Maintenance procedures should be clearly stated in the operating manual.
- Implementation must be monitored by the owner.

Ronan Point

Ronan Point was a 22-storey block of flats on the east side of London, constructed from pre-cast wall and floor units. A gas explosion occurred on the 18th floor early one morning, and because the connection between walls and floors failed, the corner wall units peeled away from the slabs. The slabs were then supported only on two adjacent sides (Fig. 6.1) and falling slabs contributed to the failure of slabs below. That is, the failure was not confined to the room where the explosion occurred, but progressed down to ground level. Fortunately the number of people killed was remarkably small.

Ronan Point was one of several similar and many other tall apartment blocks which were built to alleviate the housing shortage following the war of 1939–45. Tall buildings consisting of prefabricated components were seen as a means of accelerating the provision of housing, and the particular system had been successfully used in Denmark, albeit without installed gas. The system had been described, and the particular connection which failed had been illustrated before the accident in a paper to a conference on industrialised building at the Institution of Structural Engineers; the published

Figure 6.1 *Ronan Point – partially collapsed building.*

discussion contains no adverse comment (Ruscoff, 1966). Thirty years later, the use of factory produced components is being recommended as a desirable innovation.

For the 10 years preceding the collapse, the average number of explosions per annum in the UK, causing structural damage, was 40 (Rasbash and Stretch, 1969). Yet the building regulations contained no provision for design against gas explosion; accordingly the structure had been designed against wind suction, but not against explosion. This illustrates the mind set that can occur when designers are attuned to following regulations. It also illustrates the importance of risk assessment. A prime duty of an engineer is to ask questions; that is the essence of risk assessment. If the question 'should the effect of a gas explosion be taken into account in design', had been asked, the answer would surely have been 'yes'. But the question was never asked.

The investigation carried out at Imperial College on behalf of the designer and contractor comprised 22 different tests and was aimed partly at determining the response of the failed connection to internal pressure, and partly at checking the adequacy of other connections. A test which included vertical as well as horizontal loading is illustrated in Fig. 6.2. The tests confirmed that the failed connection had a large safety factor against vertical load and wind suction, for which the structure had been designed.

- Regulations and codes should state that it should not be assumed that all possible contingencies have been taken into account.
- Engineering courses should inculcate the necessity and art of asking questions..
- As a result of the collapse, regulations were modified to take account of gas explosions and of progressive collapse resulting from any cause. These requirements also provide a degree of robustness against unforeseen events.

Failure of a ship's deck

A tanker was proceeding under pilot's control up an estuary with a navigation channel which was maintained by dredging. The crew noticed a buckle which spread across the deck and down the ship's side. The deceleration which must have occurred when the ship momentarily touched bottom was so slight that it had not been felt by the crew. The amplitude of the deck buckle was about 600 mm (Fig. 6.3). The plating had separated from the stiffeners, which would have increased the amplitude of the deck buckle, and further reduced its resistance, so increasing the force on the side shell.

The buckling of the ship's side caused oil spillage, and pollution along the shores of the estuary. When the ship entered dry dock, local scuffing of the bottom shell confirmed that the ship had touched bottom.

The deck was stiffened by bulb flats attached by intermittent welding. It appeared that torsional buckling (tripping) had occured, and because the weld was insufficient to resist the transverse moment capacity of the stiffener web, the weld had failed and the plating was then unrestrained by the stiffeners. The connection had been adequate for normal loading but insufficient for the accidental (though not uncommon) loading which occurred.

In such circumstances it is desirable that the weld should not only allow the stiffener to reach its maximum resistance, but should ensure the integrity of the plate/stiffener combi-

Figure 6.2   *Laboratory test simulating internal pressure.*

nation beyond the point of maximum resistance. Thus it would be prudent, where exceptional loadings are possible, and where the consequences of failure are severe, to design the weld to resist the fully plastic transverse moment capacity of the web.

Fatigue cracking in ships

A class of six container ships, 213 m long, developed fatigue cracking at the ends of hatch side girders; the cracking was detected after the first voyage (Meek, *et al.*, 1971). The cracks developed further in subsequent voyages (Fig. 6.4) but the cracks had not extended into the hull girder. The cracking was attributed to the discontinuity between hatch girders, whose depth had been reduced to facilitate access across the deck. Also it was believed that the bow flare, which was intended to disperse the bow wave of the fast ships, had increased the bending moment.

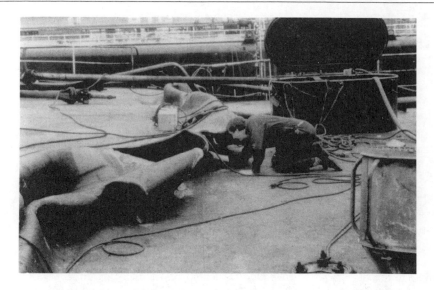

Figure 6.3 *Buckled deck of tanker.*

A rather radical solution was suggested which was accepted and implemented. The hatch girders were made continuous (so the crew had to climb over) and the ship's sides were continued upwards to a wide bracketed flange which reduced the hull bending stress, increased the stiffness and strength, and enabled additional containers to be carried, with access below.

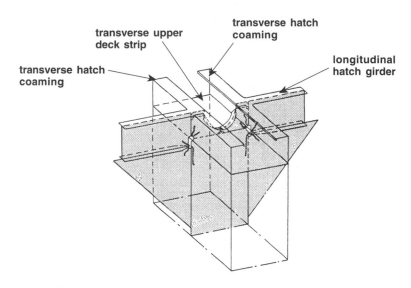

Figure 6.4 *Fatigue cracking in 213 m long container ship.*

The lessons were incorporated in the next five ships, 274 m long, and with deck strips beside the hatches only 2.78 m wide, and cross deck strips between hatches about 1.0 m wide. The beam was 32.23 m, the maximum permitted by the Panama Canal. Torsional strength and stiffness were of prime concern, and the engine space was located so that it would divide the length of 'open' deck (that is, structurally open, but the hatches had covers).

Measurements were made on scale models of the previous and of the proposed new ships and stillwater torsion experiments were conducted on one of the previous ships. Hatch corner stresses were of particular concern, and the radius was carefully chosen having regard to stress concentration and potential loss of container capacity. The stress concentration was partly offset by the use of thick insert plates. Half-scale static and fatigue tests enabled the stress distribution and the fatigue life to be estimated. A stress relieving device at the forward end of the deckhouse was also introduced. In addition strain range counting gauges were installed on the previous ships, to assess fatigue loading.

Strain gauges were fitted in regions of high bending stress and at certain hatch corners in the new ships. Strains were indicated on the bridge, and the master was given strain limits at which he would reduce speed or change course in order to reduce the strains. The facility provided comfort for all concerned, and after some time, when the precaution was found to be unnecessary, it was discontinued. I believe that this was the first occasion when strain measurements were used as a direct aid to ship operation.

The efforts made by the owners and builders to ensure safe and efficient operation of these pioneering container ships were remarkable and exemplary, as was their attitude to innovation.

Where the loading, or the structural response, is uncertain, measurement of response in service can be prudent and justified. Instrumentation in dams has been usual for many years. Quite clearly, cooperation between owner, designer, builder, and operator, is conducive to safe and efficient operation.

Collapse of a jack-up drilling barge

Sea Gem was a jack-up barge employed on drilling in the southern North Sea. The barge was moved up and down the tubular legs by a system of vertical jacks and pneumatic grippers above and below the jacks. The barge was suspended by tie bars consisting of flat bars with T-shaped ends.

Whilst the barge was being lowered, prior to changing location, a jam occurred between the legs and the wells through which the legs passed with a small clearance. When the jam cleared, the barge dropped a short distance, fracturing the T-bars on one side of the barge. The barge fell sideways and capsized. Many lives were lost, and a public inquiry followed.

A sea bed survey revealed that the legs were broken into many pieces, and were evidently brittle at the prevailing temperature. The radii at the T-ends of the tie bars were small and roughly flame-cut. In some cases the legs had torn through the plating betwccn the well and the side of the barge. The legs had a history of modification, and there were several parties to the construction, ownership and operation of the barge.

There was a presumption that when the barge dropped, the legs broke and caused the collapse. A further presumption was that if the legs had not broken, jamming would have prevented a further fall. It had not been noticed that if the drop on one side were followed by sidesway, the sway would not be arrested (by secondary jamming) until the sway was such that the leg moments would cause the stress to exceed yield stress, or alternatively the well sides would fail. Also, as some legs were strong enough to tear through the side of the well, it seemed unlikely that any legs would have failed as a result of the initial drop.

A scale model was therefore made to evaluate and to demonstrate the sway hypothesis. The legs were of ductile steel, and the platform was of solid steel to give the correct stress in the legs. The clearance between the legs and the wells and other relevant dimensions were to scale. The platform was supported by a wire on two sides, and failure of the tie bars was simulated by cutting the wire on one side of the platform. Sidesway occurred as postulated, the legs bent, and the platform fell sideways (Fig. 6.5).

Further experiments were made in which the legs were partially restrained against rotation at the bottom, to simulate qualitatively the effect of leg penetration of the seabed, but failure by sidesway still occurred.

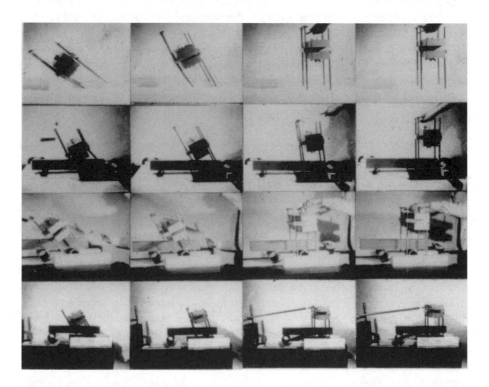

Figure 6.5   *Model used to demonstrate mode of collapse of jack-up barge.*

The model was then taken to the court room, to demonstrate the sidesway mode of collapse. There were major contractual implications, but more importantly the investigation revealed a mode of failure which would need to be taken into account in the risk assessment, in design, and in operation.

So what was the cause of the collapse? The operation did not happen as the designers presumably expected, but consistent perfection in operation should not be assumed, and the jam/drop sequence was foreseeable. Either the material specification or the supervision of testing and construction was deficient. It would appear that the mechanical, electrical and structural components of the barge were not considered as a system.

- Mobile structures, with several suppliers and previous operators, require an especially high standard of risk assessment, inspection, and operation.
- Possible departures from the intended mode of operation should be considered in the risk assessment.
- The mechanical, electrical, and structural system should be considered as a whole.
- The possibility of a mechanism developing should be considered.

Fatigue during construction

Whilst piles were being driven through the pile guides of a small accommodation jacket in the southern North Sea, a number of the welded attachments of the cathodic protection anodes failed, and many others were found to be cracked. The jacket was lifted and taken ashore for repair.

The jacket had sloping legs, and the piles were to be driven by a guided steam hammer. In the event, that hammer could not be used to its full capacity, because one of the two steam hoses was damaged, so the contractor elected to use a much heavier hydraulic hammer which was not guided, so the blows were applied eccentrically. This resulted in violent vibrations of the jacket, which applied large transverse accelerations to the anode attachments. 80% of the anodes were damaged, and several fell off. It was reported that the pile follower had a permanent deflection of 1/80 of its length and the driven end was eccentrically belled. Marine pile driving is usually subject to the exigencies of the weather, which makes construction planning and execution especially onerous.

- Installation effects can be important, and should be considered in the risk assessment.
- The designer needs to be aware of the proposed construction method, and should be notified of any change.

Collapse of a passenger walkway

The companies involved in the collapse of the Ramsgate walkway were charged and tried under criminal law, and the trial attracted much attention. The events will therefore be described in some detail.

Structures linking ships to shore must accommodate movements of the ship and the tide, whilst fulfilling the normal functions of a bridge. Whereas bridges in the UK are designed for a nominal life of 120 years, and are unlikely to become redundant in the foreseeable future, the life of a ship-to-shore structure is liable to be curtailed by changes in ship design and by the commercial vagaries of the highly competitive ports industry. Ship-to-shore structures are complex mechanical, electrical, and structural systems, yet they are subject to commercial constraints on cost.

The seaward end of a ship-to-shore structure is raised and lowered by mechanical means or by a pontoon. Berth 3 at Ramsgate originally had a vehicle linkspan supported on a pontoon; pedestrians boarded via the linkspan, which interrupted the vehicle flow. The pontoon had been designed to support an additional linkspan at a higher level, should the need arise. In due course the need did arise, and the upper linkspan was supported on two portals mounted on the pontoon. To segregate passengers from vehicles a walkway linking a new passenger building to the ship was supported by the portals (Fig. 6.6). The walkway bridge consisted of steel trusses with steel cladding, which provided an enclosed torsionally stiff structure (Fig. 6.7).

After four months in service the seaward end of the shore span of the walkway fell without warning about 10 m to the deck of the pontoon (Fig. 6.8). The violent deceleration caused the death of six passengers and seriously injured seven others.

The designers, constructors, checkers and owners were charged with offences under The Health and Safety at Work Act. Port Ramsgate was also charged under the Docks Regulations. Chapman (1998) provides a detailed account of the procurement, design and construction of the walkway, the investigation, and the legal proceedings and Chapman (1999) gives a shorter account.

Figure 6.6 *Computer generated image of the new linkspan and walkway.*

Figure 6.7 *Computer generated image of the walkway structure, seen from the shore end.*

Figure 6.8 *The collapsed walkway.*

Procurement

The existing pontoon and linkspan were designed by FKAB, a Swedish firm of naval architects and consulting engineers, which is a subsidiary of Mattson, an engineering group. They were manufactured and installed by FEAB, a shipbuilding and fabrication company, which is also a subsidiary of the Mattson group. The contract was for design-and-build and was placed with FEAB. FKAB/FEAB had also designed and built other linkspans at Ramsgate, and at other ports in the UK and elsewhere in Europe. As naval architects, consulting engineers, shipbuilders and fabricators, they were well qualified to design and construct the ship-to-shore connection. The project manager for the construction had recently returned from the Falklands, where he had been engaged on a project for the British Government.

FKAB's association with Berth 3 began in 1983, when they commissioned the Technical Research Centre of Finland and the Swedish Maritime Research Institute to report on wave conditions, pontoon motions, pontoon stability, and mooring forces, on the basis of which FKAB had designed the pontoon and lower linkspan. It was therefore natural and prudent to engage FKAB/FEAB for the upper linkspan and walkway. The walkway entered service on 12 May 1994.

The passenger building was procured under a separate design and build contract from Kayover, a local fabricator, who had already carried out a variety of tasks for Ramsgate over several years. Kayover and FEAB were asked to liaise on matters affecting the interface between walkway and building.

It was a condition of the contract that the design, fabrication, and construction would be checked by Lloyds Register, who had certified the existing pontoon and linkspan, and had performed regular surveys since commissioning.

The contract was entered in a spirit of co-operation and trust based on past association, a propitious circumstance for a successful enterprise.

Design

The walkway consisted of three spans and enabled passengers to walk from the passenger building, at about 10 m above ground, to an upper deck of the ship. The shore span extended 33 m from the passenger building to a bracket on the outside of the first portal frame (Fig. 6.9). The pontoon span extended from the first bracket to a bracket on the second portal frame. The ship span extended from the second bracket to the ship, and the ship end of the walkway was raised and lowered by hydraulic jacks. Weather protection at the span junctions was provided by flexible bellows.

The tidal range at Ramsgate is about 5 m. The moored pontoon moves approximately vertically with the tide, and is subject to wind, waves, vehicle forces, and ship impact. For the linkspans, steel on steel sliding bearings were provided at the shore end. Lubrication was applied by Greasomatics which could be actuated by chemically-induced pressure or operated manually. The vertically hinged bearings were placed under the main girders.

The walkway bearings hinged about externally projecting stub axles which were welded within a hole in a disc, which was site welded to the walkway structure. The shore bearings could slide in channels. A pad of low friction material was attached to the under-side of the bearings, and was believed by FKAB/FEAB to require no lubrication. The right hand (seen from the shore) seaward bearing was retained in position by a pintle passing through the plating of the bracket (Fig. 6.10). The left-hand seaward bearing had no pintle. A 6 mm tapping was provided on the underside of each sleeve to receive the supply tube from a Greasomatic.

Figure 6.9 *Computer generated image of the support at the seaward end of the walkway.*

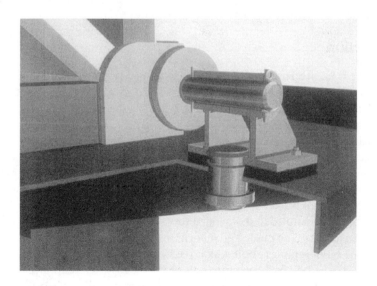

Figure 6.10 *Computer generated image of the right-hand seaward bearing.*

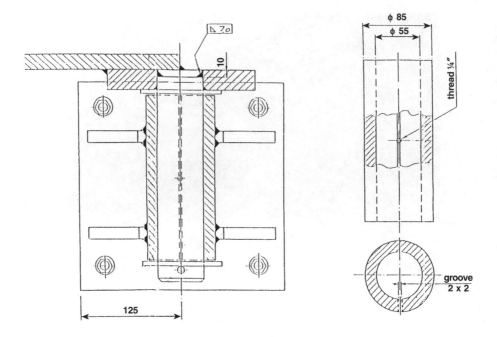

Figure 6.11   *Bearing details.*

It can be seen that of the six degrees of freedom of the pontoon, surge, sway, heave, pitch, and yaw were accommodated, but roll was not.

Construction

The bearings had been assembled at works, with the axles fixed within the bore of the discs by two circumferential welds (Fig. 6.11). The disc was site welded to the D-plates around half the circumference and on the vertical diameter.

The walkway, which weighed about 21 t, was lifted into position by crane. It was then apparent that the slideways were in the wrong position, because Kayover had assumed that the bearings would as normal be under the trusses of the walkway. It was also apparent that the walkway was too short, and might drop off the edge overnight on the falling tide. Two projecting railway sleepers were therefore lashed to the building and greased, the walkway being supported on the bottom end of the frame. During the night an appendage fouled and damaged the edge girder of the building, so imposing an abnormal force on the pintle and deforming the hole in the plate of the supporting platform. Also the retaining bolt which passed diametrically through the pintle was bent; although the bolt had been fitted, there was insufficient clearance to fit the nut. Therefore FEAB removed the bolt and welded a retaining collar. Thereafter it was only possible to remove the bearing from the platform by grinding away the weld. After several days a triangular extension 700 mm long was supplied and fitted to the shore end of the walkway, and the slideways were aligned with the bearings.

It should be noted that until the remedial work had been completed, the shore end of the walkway was supported on the end frame, and the bearings were freely suspended, but the seaward bearings were subject to horizontal forces caused by friction at the shore end, and vertical forces caused by rolling of the pontoon. That was the only effect of the above errors which was material to the collapse.

When the construction was complete, Lloyds Register conducted a load test on the walkway, as part of the installation certification. The only unexpected result was that the measured deflection was 2mm (span/16850), which although much less than expected, was recorded without comment.

Operation

When the slideways had been relocated, and the shore bearings were supporting the walkway, but the building was incomplete, the building was observed to vibrate. This was caused by juddering of the bearings on the slideways, which in accordance with FEAB's belief that lubrication was not required, had not been greased. Ramsgate, with FEAB's agreement, then greased the slideways, and the juddering ceased. The walkway then entered service. The only remarkable behaviour was that a shore bearing was observed to lift. This was at first attributed by FEAB to incorrect ballasting of the pontoon, but lifting was inevitable, as a result of the inadequate articulation.

Collapse

The collapse occurred as passengers were boarding a ferry at 00.45 a.m. on 14 September 1994. Rescue operations were very difficult (the slope of the walkway was then 1:3) and there was much suffering. The seaward end of the walkway had penetrated the pontoon deck. The shore end of the walkway rested on the edge of the passenger building, and the bottom chord was bent where it had impacted the edge girder of the building, presumably as a result of the rotation about a transverse axis caused by the impact at the seaward end. There was no other damage to the structure and the windows were not broken. The right-hand seaward bearing remained on the platform, to which it was attached by the pintle; the welds attaching the axle to the disc had failed. The left-hand seaward bearing was missing, but it was recovered by divers from the dock, whence it had been propelled by the impact. The shore bearings were still attached.

Investigation

Structural and metallurgical investigations were carried out at the Health and Safety Executive laboratories in Sheffield. Their investigations were careful and their report was thorough and objective. The dominant cause of the collapse quickly emerged from the design calculations supplied to HSE by FKAB. Nominally the resultant reactions to the vertical and horizontal forces applied to the bearing were at the centre of the bearing pads. Prudently it would have been assumed that the reactions were at the outer edges of the pads, and the bending moment on the axle, which is a maximum at its connection to the disc, and the axle and weld stresses would have been calculated accordingly. The section modulus of the axle was calculated, but it was evidently then decided that the axle bending moment at the connection to the disc was zero. This was frankly confirmed in response to HSE's questions after the accident. The axle and welded connection was only designed against shear, so the axle and connection were grossly overstressed. The pintle was in fact predominantly in shear, and it is strange that no one thought it anomalous that the axle section area was only 47% of the pintle area.

Because the axle was contained by a sleeve, which was stiffer and stronger than the axle, the sleeve alone was checked for bending. The moment was taken as the force applied at the disc multiplied by the distance from the disc to the first vertical support plate (Fig. 6.12). Apparently the designers visualised that the forces were applied without moment at the disc and that the moment was resisted by the bearing, whereas the reverse was true. A smaller, but still major, error was to assume that all reactions were equal. The explanation given to HSE was that the torsional stiffness of the structure had been thought small enough to justify that assumption.

The most extraordinary feature of the sad affair was that the checkers, who carried out independent calculations, made the same mistakes.

The metallurgical examinations established that all axles had extensive fatigue cracking. The cracking had completely severed the right-hand seaward connection (Fig. 6.13). This bearing could only tilt to the extent permitted by the pintle retaining ring, and it seems possible that the end of the axle remained within the disc for a short period before being dislodged. The left-hand seaward axle was largely cracked, but final failure occurred when the bearing struck the pontoon deck. Both shore bearings had advanced fatigue cracking. There were a number of weld defects, some of which reached the surface of the weld.

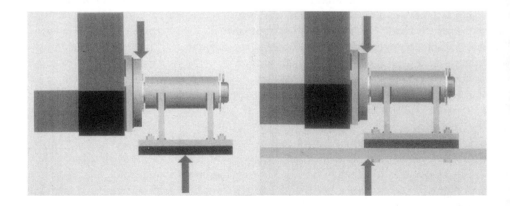

Figure 6.12 *Computer generated images of the nominal and assumed action of bearing.*

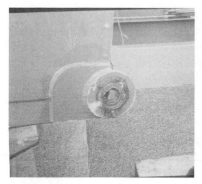 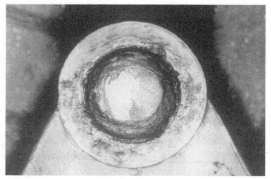

Figure 6.13 *The failed axle connection of the right-hand seaward bearing.*

Because Greasomatics had not been fitted (though Ramsgate had supplied FEAB with Greasomatics to fit where required) none of the axles had been lubricated since installation. The right-hand seaward axle and both shore axles had retained the original lubrication, and the shore axles could rotate freely within the sleeves. The right-hand seaward axle could be turned by applying a torque which would not have caused a significant increase in weld stress. The left-hand seaward axle, which had impacted the pontoon and had been immersed, could not be turned. It was removed by slitting the tube, and when examined it had no grease.

As the shore bearings had not experienced the exceptional loading which occurred during construction, and were less exposed to the weather, but still had extensive fatigue cracks, it was apparent that neither welding defects nor lack of lubrication had a significant effect on collapse, but they might have caused the collapse to occur a little sooner than it would have occurred if the welds had been perfect and if lubrication had been applied to the axles. However, as the lubrication holes were on the line of maximum pressure, it seems likely that the Greasomatics would only have been effective when lifting of a bearing occurred. If the assumption that the reactions would be equal had been correct, lifting would not have occurred. All the experts agreed that the design errors made collapse inevitable. The cause of the collapse was design error; the mode of failure was fatigue.

The bearings survived the load test, and for four months thereafter, because bending of the axles caused the reaction to move towards the discs. This was confirmed by the sliding marks on the bearing pads.

The walkway was inspected by the insurer's inspector, who was certificated to inspect ship-to-shore structures, one month before the collapse; he found nothing amiss, and of course there was nothing, apart from fatigue cracks which could only be detected by dismantling the bearings.

The trial

The trial took place at the Central Criminal Court, before Mr. Justice Clarke. FKAB, FEAB, Lloyds Register and Ramsgate were charged under the Health and Safety at

Work Act. Ramsgate was also charged under the Docks Regulations. The first six days were occupied by legal argument, and during this period Lloyds Register pleaded guilty to a lesser charge whereby they admitted error but Lloyds Rules were not impugned. FKAB and FEAB did not plead, and did not attend, so a formal plea of not guilty was entered on their behalf. Thus only Ramsgate remained in contention.

The main issue before the Court was whether under the Health & Safety Act the owner was responsible for errors made by his professional advisers, in this case the designers. The onus was on Ramsgate to show that they had taken all 'reasonably practicable' steps to ensure safety. However the prosecution had to prove that design was part of Ramsgate's 'undertaking'. The prosecution contrived to argue that Ramsgate *had* been involved in design, because the walkway had been 'tailor made' and not selected from a catalogue, so had brought design within their undertaking, and also to argue that Ramsgate had *not* been sufficiently involved and therefore had not taken all 'reasonably practicable' steps. Both arguments were of course rebutted by the defence.

Docks Regulation 7(1) states that safe means of access 'shall be provided' – that is, it imposes an absolute obligation on the owner, regardless of fault. As both charges related to the same alleged offence, Ramsgate were obliged to plead not guilty to both charges, but at the conclusion of his closing speech counsel for the defence invited the jury to convict under 7(1) of the Docks Regulations.

The judge informed the jury that it was his job to define the law, and theirs to determine facts. In the event the jury were required to decide certain points of law:

- whether Ramsgate's undertaking included design and construction.
- the intended meaning of 'reasonably practicable'. Did the jury think it meant 'not wholly practicable, practicable without undue difficulty or 'reasonable and practicable'? It would be practicable to commission five or ten independent checks, but it would not be reasonable. Clarity and reason would be satisfied if the requirement were defined as 'reasonable and practicable'.
- whether the offence lies in the existence of danger and not on the consequences, as has been authoritatively stated, and hence whether the timing of the accident was relevant to the offence. The defence argued that the risk had been there from the beginning, so timing was irrelevant. Despite submissions from the defence, the judge had refused to define 'risk'. In the sentencing proceedings the judge said he could not accept that the offence could be detached from the consequences. This implies that the extent of the offence is a matter of chance.
- was it reasonable and practicable for Ramsgate to have used a standard form of contract and to have specified proper quality assurance procedures. That would have been prudent, especially for commercial reasons. But it is extremely unlikely that that would have averted the tragedy, so the jury had to decide the extent to which non-causative acts or omissions should affect their verdict.
- the prosecution argued that if the axles had been lubricated, the collapse would have occurred later, and those particular passengers would not have been killed or injured; therefore lack of lubrication contributed to the accident. They also argued that Ramsgate should have fitted safety chains, which could have prevented the accident, even though the designers and checkers had not foreseen the need. In

his instruction to the jury when considering the provision of safety chains, the judge said 'the question is not what is reasonable, but what is reasonably practicable'.

- in their answers to the judge's questions it was apparent that the jury thought Ramsgate were culpable because they had not fitted safety chains. Perhaps this followed from their decision that design and construction were part of Ramsgate's undertaking, or perhaps they were encouraged by the judges instruction that 'the question is not what was reasonable, but what was reasonably practicable'.

Thus the jury were required to decide matters of law and opinion, as well as fact. They found all the defendants to be guilty as charged.

The judge gave the background to his decisions on sentencing. He accepted that Ramsgate's culpability was significantly different from that of the other defendants, but Ramsgate had not pleaded guilty, so was not entitled to credit on that account. Is it fair that a defendant should be penalised for defending himself? Ramsgate's case was hardly frivolous – the judge took seven hours to sum up, and the jury took eight hours to reach a verdict.

The judge also said 'By its verdict the jury had found, in my view correctly – that an owner and operator of a port cannot simply sit back and do nothing and rely on others, however expert'. They had hardly 'done nothing' – they had appointed experienced and competent designers and constructors, and checkers of world renown, which they were not obliged to do, who had performed independent calculations. He did not explain what other steps might have prevented the accident. Some interesting legal comments on the judge's conduct of the trial are included in the discussion on Chapman (2000).

The following fines and costs were imposed:

FKAB/FEAB, taken together	£1,000,000
Lloyds Register	£ 500,000
Ramsgate	£ 200,000

HSE's costs were allocated as follows:

FKAB/FEAB	£ 251,500
Lloyds Register	£ 252,500
Ramsgate	£ 219,500

Safety in ports

This project was initiated in June 1996, and the *Guide on Procurement, Operation and Maintenance* was published three years later (CIRIA, 1999). The scope is limited to linkspans and walkways. The project was supported by the ports through the Port Safety Organisation, Lloyds Register, insurers, HSE and DETR. Tenders were invited from consulting practices with relevant experience, and Posford Duvivier were selected from several strong contenders. Work began in December 1997. The cost was a small fraction of the cost of the trial. The project was guided by a steering group (chaired by the writer) whose members were drawn mainly from the funders.

A detailed questionnaire invited information from ports on number and type of structure, procurement methods, operational performance, incidents and their consequences, maintenance, training and qualifications. An analysis of the results of the survey is included in the report.

Dock structures are now subject to an array of nine acts and regulations, the requirements of which in respect of ship-to-shore structures are described in the *Guide*, as a prelude to the chapters on procurement, operation, and maintenance. The research contractor thought it prudent to retain the services of a consultancy (LACS) which provides advice on the scope and application of regulations. The machinery regulations and directives provide for a series of supporting standards; it is expected that about 600 will be required.

The *Guide* will assist owners in specifying and procuring structures and mechanical/electrical systems which are designed to be safe, reliable, and efficient in service and in ensuring that best practices in operation, maintenance, recording, and reporting are adopted. Advice is also given on procedures which must be adopted in order to conform with the regulations.

The recommendations include the formation of an industry data bank on incidents and near misses, and the drafting of a design guide and a code of practice on design of linkspans and walkways. A code drafting committee exists but so far funding has not been obtained; less than 5% of the imposed fines and costs of the trial would suffice.

Another constructive outcome of the collapse is that Lloyds Rules for linkspans and walkways have been extensively updated (Lloyds Register, 1998).

It should be noted that both these initiatives were taken as a direct result of the accident, not as a result of the trial.

- Investigators have the benefit of hindsight, and there is a tendency to think that a mistake should have been obvious to the designer. If two experienced organisations make the same mistake, then we should ask whether a chance in a million occurred, perhaps requiring no further action, or whether guidance is required to prevent a recurrence.
- Bearings or auxiliary attachments should prevent a bridge from being dislodged from the abutments under all possible loadings.
- Bearing design should facilitate maintenance, but deficient maintenance should not lead to collapse.
- Where four-point supports are provided for ship-to-shore structures, the bridge and its supports should be designed on the assumption that only two supports are effective.
- Training and qualification of operating personnel should be encouraged, in addition to equipment-specific training. This should lead to more thoughtful and responsible operation, the ability to cope with unforeseen events, improved career prospects, and greater job satisfaction.
- The first priority following an accident should be prevention.
- It appears that design will be regarded as part of a port's undertaking.

The trial also prompts some questions:

Is criminal law the appropriate instrument to punish genuine mistakes?

Is trial by jury appropriate in cases where lawyers, judges and juries have difficulty in grasping the issues? Lawyers have the benefit of specialist advice; judges and juries do not.

Should non-causative actions or inactions be taken into account in assessing guilt?

What is the meaning of 'reasonably practicable'?

Fairlead failure

A fairlead is a pedestal attached to a ship's deck around the head of which a mooring cable passes. When mooring a tanker, the welded connection of a fairlead to the deck failed suddenly; the fairlead became a projectile and caused the nylon mooring cable to impact a sharp edge, which cut the cable. The whiplash severed the lower legs of the two winchmen.

- The investigation led to the recommendation that the design rules should require that the connection to the deck should develop the strength of the fairlead, and that the deck stiffening should be arranged to transmit the concomitant forces.
- A risk assessment should recognise that mooring forces can be exceptionally large, and should consider the effect of possible failure of the rope or the attachments.

Bulkhead collapse

A tanker which had completed many voyages over eight years without incident was preparing to depart. Oil was pumped into some tanks, with adjacent tanks empty. The engine had just been started when the crew noticed that the level in one tank was falling, whilst the level in the adjacent empty tank was rising. The reason can be seen in Fig. 6.14. The departing bulkhead took with it a small portion of the ship's bottom, but the pollution was minimal, because the ship was grounded near the terminal.

Investigation showed that poor fit-up at the junction between bulkhead and side and bottom shell had been remedied by the insertion of welding rods. This did not explain why failure occurred in calm water, having survived sea-going accelerations. A possible reason was that the tank had been over-pumped, with the ventilators insufficiently open or otherwise unable to discharge freely the excess flow. However, the failure was said to have occurred two hours after completion of pumping; the crew reported that the ventilators were open, and gave no evidence of discharge from the ventilators. There was little corrosion.

The welding was certainly deficient, and there were important welding deficiencies elsewhere in the structure. Whether the failure would have occurred if the welding had been satisfactory is an open question, but the poor quality of the bulkhead welding was a major factor in the negotiation. The builders, the classification society and the owner all had inspectors on site during construction, but evidently the extent and quality of supervision did not match the magnitude of the task.

Figure 6.14   *Collapsed bulkhead of tanker.*

- Inspection reports should state areas inspected and results, during and after welding.
- Welders should be tested for the particular materials, configurations and attitudes which they will encounter.
- Fit up limits should be specified, and procedures for excesses should be established.

Fire brigade ladder

Ladders mounted on fire fighting vehicles are routinely tested by extending the ladder horizontally. On one occasion a ladder dropped to the ground, causing no injury. The ladder seemed to be an excellent piece of machinery, embodying much engineering skill and experience, but the welded steel lattice frame, through which the ladder was attached to the vehicle, was designed without proper regard to the transmission of the applied forces. Strain measurements confirmed the inadequacy. Although advanced fatigue cracking had occurred, this had not been noticed during routine inspection, because the lattice frame was encased.

- Load testing alone does not ensure safety; it should be accompanied by examination of the structure, preferably whilst the test load is applied, because cracks are then more visible.
- Engineers should have enough knowledge of other disciplines to recognise when advice is needed.
- The system should be considered as a whole.

Delayed fracture

The structures laboratory at Imperial College, London, has a prestressed floor 1 m thick. The columns for loading frames are solid circular sections 150 mm in diameter; the level of the beams is adjusted by means of grippers which can be moved up or down the columns. The gripper bolts are tensioned hydraulically. High grade high tensile bolts were installed at a presumably safe tension after confirmatory tensile testing. After periods which varied from days to weeks, some bolts fired themselves across the laboratory at lethal velocity.

Expert opinion was that the delayed fracture was caused by hydrogen embrittlement, and that the solution was to bake the bolts at a certain temperature; this, it was suggested, would expel the hydrogen. When this was found to be ineffective, the bolt holes were enlarged, and lower grade bolts were used.

Presumably the phenomenon of delayed fracture is now better understood, though more recently a manufacturer's assurance regarding bolts of the highest grade was less than absolute. Feedback of user experience would be useful.

It is important to recognise that normal tensile testing does not guard against delayed fracture and that when using extra high grade steel bolts or rods, the consequences of delayed fracture should be included in the risk assessment.

Chain failure

A space frame roof structure was being raised by chains supported by six lifting columns. Centrally controlled hydraulically actuated latches engaged in the chains and ensured that equal vertical displacements were applied at each lifting column. The force in each chain was known from the hydraulic pressure. When the roof was partly lifted, a chain link broke in two, and one piece travelled 50 m; the structure was badly damaged. The chain of the adjacent column was then carrying 2.6 times the load carried by the failed chain, but was undamaged. When the roof was being lowered a link in another chain failed whilst carrying a load smaller than that at which the first chain failed. The roof was then supported by four chains, none of which failed. The system had been successfully used on a previous occasion.

The chains passed over a pair of unfaceted grooved pulley wheels, so the links were, subjected to bending about an axis at 90° to the axis of bending caused by a straight pull. After the failure a chain was tested when passing round one of the wheels. When the force in each leg of the chain was 3.8 times the force in the first failed chain, alternate links were bent to the radius of the wheel, but the chain was otherwise undamaged. No links had been visibly bent during the lifting operation, so most links were satisfactory, but some were not. Inspection of the chain revealed that a visibly different link occurred every 5 m; the weld upset was less and the colour was different.

A tensile test of a specimen from a sound link, across the weld, was satisfactory, and failure occurred away from the weld. Metallurgical investigation showed that the welded connections of the failed links were defective and there were signs that the copper grippers which applied force and fusing current to the connection had slipped. The broken links had non ductile transverse fracture faces and there was no evidence of out-of-plane bending. The fractures were not fatigue fractures, but indicated poor fusion. There was less upset in the failed links and in some other unfailed links, which also had evidence of gripper slip; lack of upset was found to be a good indicator of deficient welds. Magnetic particle inspection of a sound length of chain revealed no flaw before or after proof load testing; links were also bent over a 38 mm radius former to a deflection of 7.5 mm; no cracking could then be detected by magnetic particle inspection. Evidently 5 m lengths of well-formed and tested chains had been joined by deficient links.

- Chain safety depends critically on weld quality, which is sensitive to small maladjustments of the automatic welding equipment.
- It is important to know who is the manufacturer (as distinct from the vendor), and to be assured that the prescribed high standards of manufacture and quality inspection are maintained.
- An inspection of the manufacturing process, quality assurance procedures and testing facilities can be instructive.
- For effective chain inspection, every link must be examined. Load testing does not assure quality unless accompanied or followed by inspection.
- A risk assessment should consider the consequences of premature failure.

Vehicle lift

The purpose of the lift was to raise vehicles from a covered deck to the weather deck of a ship. In the lowered position, a vehicle was driven up a ramp to the lift platform. In the raised position the platform acted as a hatch cover. The lift was designed to lift a vehicle weighing 45 tonnes. The platform, which was 19 m long and weighed 33 tonnes, was raised and lowered by four chains, which were similar in principle to a bicycle chain. Each had a specified breaking strength of 170 tonnes, so their nominal strength factor was large and the platform structure also was conservatively designed.

Three persons were riding on the platform for inspection purposes when the platform dropped; one person was killed and two were seriously injured. When carrying a 45-ton vehicle the nominal force in each chain would have been about 20 tons; failure occurred when the nominal force was about 8 tons.

The chains were attached to the weather deck and passed round sprocket wheels at the corners of the platform to a single hydraulic ram on each side of the platform. Two chains turned through 90° around one sprocket wheel and the other chains turned 90° around one sprocket wheel and 180° around an auxiliary sprocket wheel (Fig. 6.15).

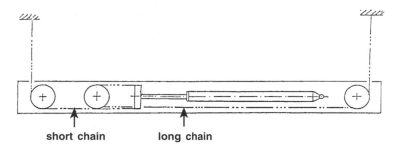

Figure 6.15 *Mechanism of vehicle lift.*

Therefore each jack applied tension to two chains of different lengths (8 m and 23 m). The chains were attached by 65 mm diameter bolts to each end of a crosshead at the end of the piston rod. The two cylinders, which were 7m long, were hydraulically connected, so the pressures were equal. To ensure that the two sides of the platform would have the same vertical displacement, the sprocket wheels at one end of the platform were connected by a shaft, the torque in which enabled the tensions in the vertical chains on each side at one end of the platform to be different.

The chain tensions were different for several reasons.

- The long chain had more friction, because it turned through 270° whereas the short chain turned through 90°.
- The long chain extended more than the short chain, so (if the slack were removed) the corners supported by the short chains would lift off the deck before the other corners. A similar situation existed on arrival at the upper deck.
- The centre of gravity of a vehicle would normally not be midway between chains, either longitudinally or transversely.
- Although in principle adjustable bolts enabled the short chain slack to be increased to compensate for relative chain extension, this was difficult to achieve even for a particular concentric loading.

Thus the tensions in the chains on each side of the platform would differ, and so would the tensions in the two chains attached to a piston rod crosshead. The crosshead was rigidly attached to the piston rod and the chains were attached to the crosshead by M60 eye bolts at 0.5 m centres. Unequal bending moments were therefore applied to each side of the crosshead and to the piston rod, which according to the evidence bent visibly under load. When the platform was in the raised position, it was supported on retractable pawls.

The piston rod was connected to the crosshead by screwing the threaded end of the piston rod into the crosshead. The rod was reduced in diameter at the threaded end, with a 2 mm radius at the reduction. For a nominally equal chain force, the axial stress in the reduced end would have been 50 N/mm^2. The stress concentration factor would have increased this to about 130 N/mm^2. The oil pressure relief valve would have limited the axial stress to about 85 N/mm². So if only axial stress were considered the highest stress which might have been envisaged would have been 40% of the specified minimum yield stress, 550 N/mm² (though subsequent investigation found that the yield stress was about half that). However, calculations of the stresses which could have occurred as a result of the various effects causing unequal chain forces, showed that the bending stress could have been several times the nominal axial stress. That was the cause of the fatigue failure.

A surviving witness said that when failure occurred there was a loud bang followed immediately by a second bang. Subsequent investigation showed that the port side piston rod had a deeper fatigue crack than the starboard side rod, so presumably the port side rod failed first.

It transpired that a few months previously piston rod failures had occurred on another ship. This prompted a recommendation that the detail should be inspected periodically. In the event that visual inspection revealed an irregularity, penetrant dye would be applied and if that confirmed cracking, a rod of new design would be requested; the logic was flawed, and the lack of urgency astonishing. It is also surprising that the operators were not alerted to danger by the nature of the recommendation.

The numerous drawings suggested care and competence. The designers displayed caution in respect of those criteria which were considered; but there was apparently little thought given to the real behaviour of the lifting system, and no recognition of stress concentrations and their effect on fatigue. The hydraulic jacks were presumably purchased items, and assumed to be of the intended standard. According to the metallurgical investigation the necessary heat treatment of the rods had been omitted, so they were deficient in strength and toughness.

- A system must be considered in all its parts, and understood in its totality.
- Where a system is not statically determinate, the consequences of small departures from nominal dimensions must be considered.
- Near misses should be reported and acted upon.
- Periodic inspection of fatigue cracks will not ensure safety where there are no alternative load paths.

Loss of the *Derbyshire*

The loss, with all hands, occurred in 1980, after four years service, during typhoon *Orchid* in the Sea of Japan, in a water depth of 4250 m, and without any distress signal being received. Bulk carrier losses are an international disgrace – there have been 108 in the last eight years (Faulkner, 1999) but in general receive little public attention, at least in the UK. *Derbyshire*, however, has been and still is a *cause célèbre,* partly because of the circumstances of the loss and partly because of the efforts of the British dependants of the crew, and of the Seamens Union, to keep alive the investigation of the tragedy.

There was circumstantial evidence that the ship might have broken in two at frame 65, just forward of the accommodation. That might have explained why no distress signal was received. Also it could suggest that the loss was caused by bad design or construction, which could affect compensation. A sister ship, *Tyne Bridge*, after nine

years service, had suffered major brittle fractures emanating from structural discontinuities below deck at the frame 65 transverse bulkhead, but had survived. Repairs were
made, using steel of grade D in place of the original grade A, and the ship served for a
further four years before going to the breakers. Other sister ships had also experienced
fatigue cracking below deck in the region of frame 65, and had been repaired, in some
instances with modifications designed to reduce the discontinuity. It should be mentioned that local fatigue cracking is not in itself exceptional.

It transpired however that the fractures in *Tyne Bridge* had occurred at about 6° C
when the ship was in ballast, which caused a tensile still water bending stress in the deck
at frame 65, whereas when the loss occurred *Derbyshire* was laden, for which condition
the still water stress at frame 65 was compressive. Also the reported air and sea temperatures were 28° C and 27° C, and according to Lloyds Register records of 40 brittle
fractures, the highest temperature at which fracture had occurred was 24° C, the next
highest was 19° C, and the average for the 40 ships was 3° C. In the writer's opinion
therefore it seemed unlikely that catastrophic failure at frame 65 had caused the loss.

Since that initial opinion, underwater surveys have been undertaken, the most recent being extensive and detailed. The results, and associated analysis of 14 possible
causes of failure, are described by Faulkner (1999). The paper, which considers hydrodynamic and structural effects, is authoritative and detailed, and should be read by
anyone wishing to examine possible causes. Many recommendations for improving the
safety of bulk carriers are given.

Faulkner concludes that because implosion occurred on both sides of the bulkhead
at frame 65, and because any compartment which imploded must have been intact at
the time of sinking, deck fracture at frame 65 can be ruled out as the cause of the loss. It
is contended that wave forces on hatch covers during the typhoon could have been much
in excess of the strength of the covers, and that:

> *Beyond any reasonable doubt, the direct cause of the loss of the m.v.* Derbyshire
> *was the quite inadequate strength of her cargo hatch covers to withstand the
> forces of typhoon* Orchid. *This weakness to resist substantial water ingress is
> gross when compared with other major elements of the watertight boundaries of
> the ship's hull.*

> *These hatch covers did meet the acceptable stress criterion of the 1966 Internation-
> al Convention on Load Lines. It then follows that the fundamental fault and cause
> of this tragic loss lies fairly and squarely in the altogether inadequate value and
> inappropriate nature of the loading and safety factor implicit in these Rules.*

Because of the fragmented nature of the wreckage, it seems unlikely that the cause
or causes of the loss will ever be established with certainty. That does not diminish the
importance of the investigations which have continued over 20 years. If causes were
possible, then they should be addressed in design.

The conclusion that deck failure at frame 65 did not cause the loss, does not imply
that fatigue cracking below deck, which occurred in all six ships of the class, is acceptable. Neither does it imply that the use of grade A steel in primary structure is acceptable. Load specifications for hatch covers (which are primary structures inasmuch as

watertightness depends on them) have been increased by some classification societies since 1980, but larger horizontal forces than are currently considered can be exerted on hatch coamings. Because of the wide variation of short duration peak pressures, it is important that coamings should not be prone to brittle fracture under impact; they also are a primary structure.

The investigation by Skinner (1987) on liquefaction of unpelletised iron ore, whether caused by progressive increase in pore pressure due to ship motion, or by water ingress, has demonstrated the desirability of conveying iron ore in pelletised form. Four bulk carriers are known to have experienced major cargo shift; three were lost, one was towed to port. The phenomenon should not be dismissed because it caused only 1.1% of losses in the period 1960–94. In the period 1969–87, 128 bulk carriers were lost, that is seven per annum. In the period 1991–1999, 108 bulk carriers, of average age 19 years, were lost, that is 13 per annum. However it appears that about 35% were caused by navigational errors (Faulkner, 1999). Fire and/or explosion was another important cause. Structural failure is only one of several causes that require action to improve safety (there is no suggestion that crew performance contributed to the loss of *Derbyshire*). The age of ships lost suggests that corrosion could be an important factor. The increase in annual casualty rates since 1991 suggests that if lessons have been learned, they have not been applied.

Faulkner in particular and the many other authors listed are to be congratulated for the sustained effort over 20 years to improve understanding and knowledge. By December 1999, the parties to the Formal Investigation had been asked by Mr. Justice Colman to provide provisional answers to a list of questions (RINA, 1999).

Whilst research is an ongoing requirement, it seems to the writer that the most urgent problem is not lack of knowledge, but lack of implementation and therefore
 • marine losses should be more widely publicised.
 • discontinuities should be minimised, and their effect taken into account.
 • grade A steel in primary structure should not be permitted.
 • shipping is international and competitive, so stringent international regulations, accompanied by enforcement, are essential for safety.
 • condition surveys should be regular, competent, and acted upon.
 • where relevant, the effects of possible liquefaction should be considered.

Early warnings

The effects of contaminated aggregates on concrete durability are now well known, but the rapid deterioration of some structures in the Middle East in the 1970s can be used to emphasise that the effects of exceeding specified limits of chlorides and sulphates are real, and can have major economic consequences.

A straight starter bar broke when a small deviation was being rectified, casting doubt on the ductility of all similar bars on the site. Was this a unique occurrence? Are site bend tests advisable?

The driving record for a precast concrete pile indicated abnormality, and this was confirmed ultrasonically. Excavation showed that shear failure had caused the reinforcement to develop an S-bend. A central hole provides a simple means of checking integrity.

None of the above events created a safety risk. The first had major cost implications for clients and contractors. The second caused disruption to programme, and possibly was a unique event. The third emphasised the need for careful driving records and detailed inspection, assisted perhaps by automatic plotting. A central hole facilitates quick and convincing checking.

Lessons and actions

- With diligence and vigilance, safety and quality may be maintained without regulations. Regulations can be helpful, but without diligence and vigilance they will not be effective.
- The essence of risk assessment is that the designer considers the effects of all possible eventualities, as distinct from checking that specified loads can be supported assuming that materials perform as expected. Risk assessment can be assisted by a check list compiled with the help of experience. However, it should be made clear that the list is intended to stimulate and not to inhibit the imagination.
- Specialisation can lead to partition of responsibility. An overall view of parts and effects is required. For example the safe and reliable operation of linkspans and walkways in ports depends on structural, mechanical and electrical components, as well as on aspects of ship operation. The CIRIA report (1999) therefore recommends the designation of a system engineer who will appraise the design, operation and maintenance of the whole system.
- Experience should be fed back promptly to design practice, to codes, and to engineering education.
- Engineering courses should include an obligatory module devoted specifically to safety and performance. The module should include an introduction to an engineer's moral and legal responsibilities in design, construction and operation, and to risk assessment and analysis.
- Courses should at least make students aware of the questions which need to be asked regarding the properties and performance of construction materials.
- Engineering courses should ensure that computer aided analysis, and design codes, are not seen as a substitute for physical grasp.
- Provision for inspection, maintenance and safe operation should be included in design, and clear instructions should be issued. The effects of deficient maintenance or operation should be considered in the risk assessment.
- In the case of movable structures, operator training and qualification is essential.
- In many circumstances it is necessary to ensure that connections should be able to develop the strength of the parts joined. Where this is considered to be unnecessary, the justification should be recorded.
- The circumstances under which a structure could become a mechanism should considered.

- The designer should have evidence that material properties are as specified. The identity of the manufacturer (as distinct from the vendor) should be known.
- Lifting structures and moving structures demand an especially high standard of design, procurement, operation and maintenance.
- For multi-discipline systems, responsibility for interaction between components, and for coordination of disciplines, should be clearly allocated.
- Regulations are now so numerous that consultants are being employed solely to advise on regulations. For example ship-to-shore structures are subject to nine sets of regulations. Each has its virtues, but digesting and fulfilling the formal requirements of the totality might demand disproportionate effort, to the detriment of design, analysis, checking and supervision. Guidance which is specific to particular types of structure, including a distillation of relevant regulatory requirements, can help to realise the intention, whilst redressing the disproportion.
- Engineers should be aware of the danger of their imaginations being inhibited by regulations and design codes, which should make clear that they do not cover every possible eventuality. The designer, constructor and operator should see their prime objectives as being to ensure safe and efficient performance of the structure or machine. They should then check that the regulations have been satisfied and that any relaxation of code requirements can be justified.
- Accidents have been caused by an incorrect visualisation of the action of a component or system, by omitting to consider a certain phenomenon, or by not taking account of some operational circumstance. Before embarking on calculations, it is important to write down the loadings and phenomena to be considered, and the idealisations to be made in the analytical model. Reference would be made to the risk assessment.
- Connections are of special importance. They usually determine fatigue life and the mode of fatigue failure. If they are at least as strong as the adjoining parts they will greatly enhance the robustness of the structure in the event of extreme or unforeseen loading. If the integrity of the components is maintained, the consequences of collapse will be less severe. Connections also have a major impact on cost.
- When designing against fatigue, the redundancy of the structure should be considered. If alternative force paths exist, crack monitoring might be an acceptable means of delaying repair or strengthening. If there is no alternative force path, then failure follows quickly after crack detection, and immediate action is required. Therefore it is logical in design or assessment to allow a greater margin against calculated fatigue life when there is no alternative force path.
- The reporting of minor incidents and near misses is very important. If a major disaster occurs, then action to prevent a recurrence will probably be taken. If the minor incidents and near misses are reported and disseminated, major disasters could be prevented.
- Notwithstanding the proper emphasis on safety, it should be remembered that even failures which are not life threatening, can have major economic consequences. Safety and economy are compatible.

• A system is required for collecting, collating, and disseminating information on failures and near misses. The scope would need careful consideration. For moveable port structures, a reporting system is now in operation.

References

Chapman, J.C. 1998. Collapse of the Ramsgate Walkway. *The Structural Engineer*, **76**, 1, pp. 1–00, and discussion, *The Structural Engineer*, **77**, pp. 22–29.

Chapman, J.C. 1999. Lessons from the Collapse of the Ramsgate Walkway. *ICE Conference Forensic Engineering*, Thomas Telford.

Chapman, J.C. 2000. Learning from failures. *The Structural Engineer*, **78**, 9, pp. 23–31.

CIRIA 1999. Safety in Ports – Ship-to-Shore Linkspans and Walkways – A guide to procurement, operation and maintenence. *CIRIA Report* C518.

Faulkner, D. 1999. *An analytical assessment of the sinking of the M. V.* Derbyshire. RINA W218.

Lloyd's Register 1998. *Rules and Regulations for the Classification of Linkspans*. Lloyd's Register of Shipping.

Meek, M., Adams, R., Chapman, J.C. and Reibel, H. 1971. The Structural Design of the OCL Container Ships. *Transactions RINA*, **114**, pp. 241–242

Rasbash, D.J. and Stretch, K.L. 1969. The relief of gas and vapour explosions in domestic structures. *The Structural Engineer*, **47**, 10.

RINA 1999. MV *Derbyshire* – Formal Investigation'. *RINA Affairs*, October.

Ruscoff, B.B. 1966. *Industrialised Building and the Structural Engineer*. IStructE Conference.

Skinner, A.E. 1987. *Cargo movement due to forces acting on it in a seaway*. Part of LR9, Lloyd's Register of Shipping.

Tietz, S.B. 1988. Risk Analysis – Uses and Abuses. *The Structural Engineer,* **76**, 20, pp. 395–401.

Acknowledgement

Some of the incidents described were investigated by Chapman & Dowling Associates and came our way through the reputation of my colleague, friend and erstwhile partner, Professor Patrick Dowling CBE DL FREng FRS.

7 Risk management from the lawyer's point of view

Diana Holtham

Introduction

What is risk management?

The vast majority of building failures are not dramatic collapses but far less newsworthy – structural or mechanical problems and issues of serviceability such as minor cracking and water ingress. These are still defined as failures both by the building owner and all those involved in the inevitable claims for compensation. So too are disputes about delays and cost overruns which turn what could have been a successful project, into a legal minefield.

Risk management, from the contractor's or designer's point of view is concerned with evaluating technical performance, but from the lawyer's point of view, it consists primarily of understanding and dealing with the risks which might give rise to a liability to pay damages. This involves ensuring that the contractual obligations undertaken are not too onerous, that they are well understood by everyone involved in the project, that suitable records are kept and that disputes, when they arise, are dealt with efficiently and effectively.

Claims trends

Claims against design professionals took off in the 1970s. Since then the numbers of claims have been affected by the boom and bust cycles in the construction industry. It is to be expected that more claims will originate in periods when construction activity is intense and everyone is busy. This is partly the result of the sheer numbers of construction projects and partly because at those times everyone is under greater pressure. There is often some delay between the problem arising and being discovered and a further delay before legal proceedings are commenced, so that the litigation trend tends to follow at least a couple of years behind the ups and downs of the construction industry.

There is though an opposite trend occurring when the economy is in poor shape. At this time clients can be more determined to obtain compensation and exact retribution for any perceived difficulties with their construction projects, particularly in relation to delays (time is money) and budget overruns. Failures in the construction industry do not just consist of physical defects. Indeed, over recent years, there has been a trend away from claims based on technical problems and defective buildings, to claims based on financial losses caused by poor cost estimating and delays to the work once the project is on site.

There has also been an increase in claims arising when clients, whether genuinely dissatisfied or not, fail to pay their advisor's fees and meet any threat of a claim for those fees with a counterclaim which can greatly exceed the fees in dispute and involve much time-consuming and expensive investigation if it is to be refuted.

Need for insurance

Professionals would be most unwise to practice in this litigious age without professional indemnity insurance. Over the years, insurers have responded to the increased tendency to sue by increasing rates and excluding or limiting cover for certain risks by careful drafting of the terms of the policy. This applies particularly to perceived new risks, the full consequences of which cannot easily be predicted, such as pollution and environmental contamination claims, an area in which statutory liabilities have increased significantly. Recently insurers have been coming to terms with the introduction of adjudication for construction disputes which has significant implications for their cash flow since it results in the much speedier resolution of disputes and claims, and the Contracts (Rights of Third Parties) Act which can give third parties with an interest in a project the right to sue without the need for a collateral warranty. It remains to be seen whether these measures will have a significant effect on insurers' exposure and how they will react if this is the case.

Claims feedback

It might be thought that insurers, who see so many examples of failures in the construction industry, would be in an ideal position to feed back information to the industry to help improve performance. However, it is an insurer's duty to defend claims against the insured to the maximum extent possible, not to publish the insured's problems and failures. In any event, apart from the occasional technical problem, such as high alumina cement which received a great deal of publicity and did change working practices, the majority of problems arise from far more mundane causes.

Most mistakes are not new, but arise from the same causes time and time again. It is quite common for lawyers who have defended a practice when a mistake has been made by one partner or director, to find themselves dealing with another claim arising from a similar mistake by another member of the firm a few months later. This is because the firm has not learned the lesson and changed its procedures appropriately. It can be extremely difficult to get this message across even to the individuals working within one practice, let alone the industry as a whole.

If lessons are to be learned from past failures, the industry requires a greater sense of openness which will be difficult to achieve when liability for substantial sums of money is at stake. Project insurance, or at the very least speedier resolution of legal disputes, so that an independent and unbiased examination of the issues can be made promptly, would help to bring problems out into the open. The majority of errors however, can be explained by poor management and administration.

Whilst openness between members of the team is likely to improve relationships and avoid problems occurring, or at least getting out of control, some forms of partnering can actually encourage claims. Partnering is still a relatively new concept, but if everyone involved in the project is expected to bare their souls, there is a risk that confessions will be used against those making them if conflict is not successfully avoided. Similarly, it is not a good idea to create internal documents explaining what has gone wrong on a project because these are potentially disclosable in any subsequent proceedings. Quality assurance schemes which require too much to be documented in this way need to be reviewed carefully.

Some useful information can be derived from claims statistics and I am grateful to Griffiths & Armour, brokers to the Association of Consulting Engineers, for information regarding their analysis of the causes of civil engineering, structural engineering and services claims notified to them by members of the ACE over a 10 year period. Whilst failure in design concept is the category accounting for the largest number of claims, such problems still amounted to substantially less than half of the claims recorded. Even within that category many problems result from delegation without adequate checking by experienced engineers rather than the inherent difficulty of the project.

Pre-contract

Risk reduction

The arrival of a new contract in the office is a time for enthusiasm and perhaps excitement. The last thing a contractor or designer wants to do is put obstacles in the way of his own appointment, but he should be careful not to rush over-optimistically into the unknown. He needs to consider carefully what he is being asked to do, whether others involved in the project are likely to be able to fulfil their roles adequately and whether the risks inherent in the project have been reasonably allocated. He should remember that he may have liabilities to others apart from his client because of statutory duties, liabilities in tort and collateral warranties. Such duties can be less easily regulated than those based upon a formal written contract negotiated between the parties. He will need to ensure that he has appropriate insurance with a sufficient limit of indemnity, a manageable excess and no unreasonable terms or conditions.

Eliminating risk is both impossible and commercially undesirable, but limiting and defining risk is essential. Risk awareness involves understanding how claims can arise and controlling them so far as possible. Adequate quality control and risk management procedures must be put in place. In these days of quality assurance procedures, such as ISO 9001, it is probable that these procedures will be set out in a manual, but it is essential that quality assurance should not degenerate into an unwanted layer of bureaucracy. All staff must understand the reasons for the procedures which they are asked to implement and this requires a process of education and communication within the practice and a willingness to discuss past mistakes and the lessons to be learned from them. Analysing what has gone wrong and feeding the results back into a practice's management systems should be a priority.

Unless it is written purely for the benefit of those who sign contracts on your practice's behalf, there is no need for a risk management manual to try to turn your staff into lawyers. Instead, it should provide practical guidance on situations likely to give rise to claims, such as the examples given in this chapter, and set out the procedures the practice wishes its staff to adopt when warning signs are detected or complaints are made. This ensures that problems are resolved effectively and claims are avoided or dealt with efficiently. All members of staff should be aware of the limit of their authority and should be encouraged to report potential problems to their project director or equivalent without fear of blame or worse.

Special risks

Different types of contract and project have their own special risks which need to be understood and taken into account. For example:

- In PFI (private finance initiative) projects enormous financial risks, associated not only with the construction of the project but its subsequent operation, may well be placed upon the contractor. The contractor will in turn try to pass these risks to sub-contractors and consultants. This can result in the smallest members of the team, with the least financial resources, being made responsible for guaranteeing performance on time and on budget when the consequences of failure could be far beyond their ability to pay, whatever level of insurance cover they carry.

- Design and build contractors are also keen to pass on the financial consequences of under-pricing a job, costed at a preliminary design stage, to designers who could not possibly have foreseen all the complexities of the job at that early stage.

- Complex projects, which will take a number of years to complete, are often subject to substantial client changes with dramatic consequences for budget and timetable, yet clients may not understand the consequences of their changes of heart or may seek to contract on terms which put the risk on the contractor or design team.

- Even in a conventional project where substantial sums may be at stake, it is possible that a member of the team will go into liquidation or be unable to fulfil its financial responsibilities in full, leaving another member of the team to pick up the liability on a joint and several basis. Before entering a contract it is important to assess the capacity of other members of the team to fulfil their responsibilities, financial and otherwise. Contractual terms can be incorporated which limit any contribution to a joint liability to a fair proportion, whether or not the others who share responsibility have paid their proportions.

Insurance for professionals

Protection against risks which cannot be eliminated can, in many cases, be obtained by appropriate insurance. Professional indemnity policies are based upon establishment of legal liability, e.g.

> *The Insurers will indemnify (the professional) in respect of loss arising from any claim made upon him during (the period of cover) in respect of any legal liability or alleged legal liability arising out of the conduct of the practice as (design professionals).*

The purpose of the policy is primarily to protect the insured, but increasingly clients insist upon knowing that the policy exists and specifying the minimum level of cover. Some contractual terms seek to prevent the professional and his insurer reaching agreement about the settlement of the insurer's liability without the consent of the professional's client. Such terms should be resisted wherever possible, although commercial considerations may make this impossible.

Professional indemnity insurance should be taken out for an adequate sum with reputable insurers and the insured should be very careful to comply strictly with the terms of the policy, particularly with regard to notification of all circumstances likely to give rise to a claim. Recently some insurers have introduced stringent terms in relation to notification. Where an adjudication is to take place, it is essential that such terms are complied with in view of the very short time scale allowed for the adjudicator to reach a decision. Failure to do so could mean that an insurer's position is seriously prejudiced and that, in turn, would make it likely that insurers would refuse to indemnify.

Insurance for contractors

Insurance policies available to contractors are more limited in scope than professional indemnity policies. Contractors' liability policies are normally intended to provide cover for injury to persons and damage to property, including the project itself, but not for poor workmanship whereas a professional is insured against liability for poor design work.

Contractors can also buy professional indemnity insurance, and any contractor who undertakes a measure of design, including specialist sub-contractors and design and build contractors, would be very unwise to do otherwise. A particular feature that a contractor needs in his professional indemnity cover is that the loss will often emerge as an extra cost he is contractually obliged to bear, rather than a liability to a third party. He therefore needs to ensure this is covered.

Space does not permit a full examination of the insurance requirements set out in the standard forms of building contract, but the liability clauses and insurance clauses are intended to work together to ensure that both contractor and employer are protected. For example, clause 20.1 of JCT 80 makes the contractor liable for claims arising out of the personal injury or death of any person caused by the works, and clause 21.1 requires insurance to cover this liability. Clause 20.2 deals with the contractor's liability for loss or damage to property (excluding the works themselves and materials delivered to site). The contractor is responsible for loss and damage arising in the course of carrying out the works if he has been negligent and this liability is also required by clause 21.1 to be covered by insurance.

Insurance of the works is dealt with in clause 22 which sets out various options. This is the so-called Contractors All Risks policy (which certainly does not cover all risks). It does in fact cover physical loss or damage to the works executed and materials delivered to site, but excludes defects due to wear and tear, damage caused by a defect in design, or by poor workmanship and damage caused by war, insurrection, etc.

Latent defect insurance

Latent defect or project insurance, providing it contains a waiver of subrogation rights, protects the whole team including contractors and consultants against liability for defects which existed at the date of Practical Completion but had not, by then, been discovered. Unfortunately it is common for clients to consider that the expense of the premium can be avoided by relying upon the contractors' and professionals' own insurance policies, but that can lead to litigation to establish a legal liability and to contribution claims between members of the team which would be avoided if they were all covered by one policy.

The contract

Contract forms

A contract, in this context, is an agreement to perform work in return for payment. Contracts can be incorporated in a formal contract document such as a standard form of building contract or consultants appointment or less formally in a simple letter. Unwritten gentlemen's agreements however are best avoided since the gentlemanly virtues tend to disappear rapidly once a dispute arises over a substantial sum of money.

There is a building contract or professional appointment form suitable for most situations. Such contracts should be amended only with care and legal advice, since it is dangerous to look simply at the meaning of one clause without understanding how it may inter-relate with other clauses. On the other hand, contracts should not simply be taken off the shelf and signed without ensuring that they are entirely appropriate to the duties being undertaken on a specific project.

Common points to look for in professional appointments include:

Duty of care

A professional's normal common law duty is to carry out his responsibilities using reasonable skill, care and diligence. He does not, and should not, guarantee performance, and fitness for purpose clauses should be avoided, not least because they are uninsurable.

Duties

Consider carefully, ambiguity must be avoided. Each party needs to understand what he is supposed to be doing, what other parties have agreed to do and where the boundaries lie. They need to consider who is responsible for coordinating the activities of the different members of the team. Consider listing additional duties which may or may not be required but, if required, will carry an entitlement to additional fees.

Budgets and programmes

These should be realistic and professionals should avoid undertaking to comply with either, since the financial consequences of a failure to comply could, and sometimes do, exceed the cost of constructing the building in question. Architects and engineers may be experts in building design, and contractors in building construction, but they are not necessarily so familiar with the world of finance agreements.

Supervision

Designers should be wary of agreeing to supervise construction unless the meaning of the term supervision is clearly defined. In general, the term inspection is to be preferred, but even this should be qualified to indicate the regularity of the visits to be made to site. Frequently this is left to the designer to determine, e.g.

> *The (designer) shall make such visits to site as he shall consider necessary to satisfy himself that the Works are executed generally according to the contract and otherwise in accordance with good engineering practice.*

Redesign

Who is to pay for redesign? A fee based on the cost of the works or a fixed fee will not compensate the designer for additional work necessitated by client changes or cost reduction exercises. Watch out for clauses requiring the designer to obtain client approval before carrying out additional work. It is easy to forget about this and find that you are not entitled to payment as a result.

Payment

Have the fee schedules in the contract been completed? Are the payment intervals reasonable? Is interest recoverable on late payment? Does a fee based on the cost of the works cover preliminary advice? What fee would be payable if the client decided not to proceed?

Advice

There may be an obligation to advise on the appointment of specialist sub-consultants, whether tests and site investigations are required, what level of site supervision is appropriate or what form of contract should be used. Take care not to go outside your field of expertise in relation to any of these matters.

Sub-consultants

Is the lead consultant to engage others as sub-consultants or will they be engaged by the client? Remember that, if sub-consultants are used, the lead consultant will be responsible both for their fees and their mistakes, if any.

Common points to look for in construction contracts are:

Liability for injury to persons and damage to property

Understand both your liabilities and the related insurance obligations (*see Insurance for contractors*) and make sure the insurance is arranged.

Design liability

Contractors can be responsible for specialist design and for selection of equipment and materials to meet a performance specification. Note that approval by the client or his design team does not relieve the contractor of his responsibility.

Site access

Check the arrangements for obtaining possession of the site, and obtaining access, and consider the problems sharing the site with any other contractors can cause in relation to working space. Permitted working hours must also be taken into account. Evaluate the risk of damage to adjoining premises and noise nuisance.

Completion date

Is the completion date and the programme realistic and achievable? Do the designers know when their details will be required? Is there sufficient time to place orders and obtain delivery of materials and equipment?

Solvency

Choose your domestic sub-contractors carefully, both from the point of view of competence and financial standing. Make sure the sub-contract is back to back with the main contract and provides suitable indemnities to cover any liability arising out of the sub-contractor's failure to perform.

Collateral warranties

It is common for the contract to require the designer or contractor to enter into collateral warranties in favour of other parties, notably future purchasers or tenants of the building, and funding bodies. A warranty is a contract in itself giving the beneficiary a direct right to sue for breach. If there is no contractual duty to give a warranty and no compelling commercial reason for doing so, they are best avoided. If they must be signed, the terms of the warranty must be considered with care and in no circumstances should the duty undertaken to the warrantee be any more onerous than that undertaken to the original contracting party.

Common points to look for in a collateral warranty include:

Duty of care

For professionals this should be limited to reasonable skill and care, although there is often a reference to experience of projects of similar complexity, e.g.

> *The Consultant has exercised and will continue to exercise all the reasonable skill care and diligence to be expected of a competent (architect/engineer, etc.) experienced in carrying out work of a similar size, scope and complexity to the Development.*

Joint and several liability

This can be excluded by a clause such as:

> *The Consultant's liability under this Clause shall be limited to that amount which it would be fair and equitable to require the Consultant to pay having regard to the extent of the Consultant's responsibility for the same and on the basis that the Contractor and all other professional consultants appointed by the Client in connection with the design and/or construction of the Development shall be deemed to have provided contractual undertakings to the beneficiary in respect of the performance of their services in connection with the Development and shall be deemed to have paid to the beneficiary such proportion as it would be just and equitable for them to pay having regard to the extent of their responsibilities.*

Heads of damage recoverable

It is sensible to limit liability to the cost of repairs since consequential losses suffered by a future occupier will depend upon the nature of his business and may therefore be unpredictable at the time the building is being designed and constructed, e.g.

> *In the event of any breach of its obligations the Consultant's liability shall be limited to the reasonable cost of repair renewal and/or reinstatement of any part or parts of the Development to the extent that the Beneficiary incurs such costs and/or becomes liable either directly or by way of financial contribution for such costs.*

Assignment

It is reasonable for the beneficiary to have the right to assign the benefit of the warranty, but to avoid finding that numerous people have acquired a right of action, it is preferable to limit the number of assignments and prevent an assignment to someone acquiring only a partial interest in the building, e.g.,

The beneficiary may without the consent of the Consultant assign any benefit arising under or out of this warranty to a person acquiring the whole of the Beneficiary's interest in the Development.

Limitation

This should not exceed the limitation period applicable under the contract between the consultant and his client so do not sign a warranty in the form of a deed if the contract is not a deed. Care should be taken when a warranty is being signed some time after Practical Completion, that the limitation period, whether 6 years or 12 years, runs from Practical Completion and not from the date the warranty is signed.

Step in rights

These may be required to enable a funding body to take over the client's rights and obligations in order to protect their investment, should the client/developer go into liquidation. The right of the consultant to determine the contract for breach by the client will be restricted if the funder is willing to assume all the obligations of the client, including the obligation to pay the consultant's fees, e.g.

The Consultant warrants and undertakes to the beneficiary that the Consultant will not exercise or seek to exercise any right of determination of the (Consultant's Appointment) or to discontinue the performance of any of the Consultant's obligations in relation to the Development by reason of the breach of the (Consultants Appointment) on the part of the Employer without giving to the beneficiary not less than 28 days' notice of the Consultant's intention to do so and specifying the grounds for the proposed termination.

Insurance

It is common for the minimum level of cover to be specified and for the beneficiary to be given the right to inspect documentary evidence that insurance is being maintained. The period during which the consultant is required to maintain insurance, if specified, should be related to the limitation period, i.e., not more than 6 years or 12 years if the warranty is in the form of a deed.

The Contracts (Third Party Rights) Act

The Contracts (Third Party Rights) Act 1999 gives a third party with an interest in a construction project the right to enforce the contract, if the contract expressly provides that he may do so or if it purports to confer a benefit on him. There is clearly considerable scope for disagreement about whether, and to what extent, a contract purports to confer a benefit on a third party. Fortunately, the act is facilitative in the sense that if it is made clear in the contract that no third party rights are intended to be conferred, none will be unwittingly created. The ACE conditions provide that *Nothing in this Agreement confers or purports to confer on any third party any benefit or any right to enforce any term of this Agreement.* A similar term should be incorporated in any other form of contract which is adopted unless it is intended to confer the right to enforce the contract on anyone who is not a party to it, in which case that person or persons should be clearly identified. Until such time as the industry has sorted out standard and acceptable clauses and these have been tested in court, it is probably sensible to seek legal advice on the drafting of any such clause.

During the works

Recurring themes

Although the circumstances of each project are inevitably different, there are a number of recurring situations which give rise to claims. Below are some examples.

Making the client's decisions for him

The budget was limited, so the engineer decided to economise on the site investigation by limiting the number of trial pits and boreholes. He did not explain the risks to his client but simply assumed that, had the risks been explained, the client would have been prepared to accept them. The construction of strip footings commenced before the true nature of the site was revealed and then there were delays and additional costs before a piled foundation could be substituted. The client would have had to pay for a piled foundation if properly advised at the outset, but the change from strip footings to piles changed the whole financial basis for the project and resulted in the client losing confidence in his designer. The risks of economising on the site investigation should have been explained and the client should have made the decision whether or not to accept the risk.

Design criteria

The parties had worked together in the past and were good friends. The designer discussed the project with his client over a drink or two and literally made notes on the back of an envelope. The client said that a structure which would have a limited life would be adequate for his purposes and that is what was built for him. When it failed a few years later the client could not recall any discussion about the design life and the engineer could not prove that it had taken place. The design criteria should have been agreed at the outset and put in writing, particularly as the client was agreeing to a standard below that which most clients would normally expect.

Design and build

The designer produced a preliminary design for the client and once a contractor was appointed he was novated to work for that contractor and asked to develop the design further. As the design developed the project became more expensive, but the contractor had tendered on the assumption that the preliminary design was feasible and alleged that the engineer's preliminary design was defective, rather than merely preliminary. The contractor should have been made aware of the preliminary nature of the design, preferably before he tendered, but certainly before the designer agreed to work for the contractor.

Client changes

On a complicated construction project, a mechanical and electrical engineer required the architect's and structural engineer's drawings before he could produce his own completed design. The client ignored the contractual requirement for a design freeze and continued requesting changes which were implemented by the architects and structural engineer. The effect of this was to seriously delay the M&E engineer's preparation of his drawings, which in turn delayed the project, with the result that the employer had to pay significant additional sums to the contractor. The architect, as team leader, and the project manager should have controlled and educated the client and insisted upon a design freeze. The engineer should have made it quite clear that he was being delayed by others.

Reliance upon information supplied

An engineer carrying out a site investigation was given information about services on the site, compiled by the client's services engineer. The presence of a telecommunication cable was not indicated on the plan supplied to the engineer, and this was struck by the drilling rig. The engineer had not recorded in writing that he was relying upon the information supplied to him and that he would not go through the exercise of checking with each of the relevant utilities as this had already been carried out earlier by others.

Limited brief

An engineer was requested to supply calculations for a domestic extension for a very small fee. The client did not want him to incur the expense of visiting site or inspecting the contractor's work. Problems arose because of the proximity of trees and the contractor's failure to take the foundations down to an adequate depth. The engineer should either have insisted upon visiting the site, refused the commission if this was not permitted, or have given the client a written disclaimer.

An engineer was asked to upgrade the equipment in a boiler room and expressed the view that the whole distribution system should be examined. The client could not afford to upgrade the whole system and so works were carried out in the boiler room alone which, due to the antiquated system outside the boiler room, did not function properly. The engineer was blamed for failing to provide a fully functioning system. The engineer should not have accepted a limited brief without warning specifically about the risks.

Failures of coordination and management

A large firm of consulting engineers operating in a specialist field and employing a range of experts with particular specialities, designed a commercial plant which proved to have inadequate capacity. No one person was prepared to take full responsibility for the coordination of the work of the different experts, either during the project or after problems arose and litigation commenced. The case had to be settled, rather than risk witnesses blaming one another. Proper project management might have ensured the plant operated properly and would certainly have strengthened the firm's position in the litigation.

Gratuitous advice

An architect was having difficulty deciding what fixing system he should specify for some cladding which, although non-structural, was to be applied to a structural element of a building, namely the concrete frame. The structural engineer was drawn into the debate and made a suggestion which was adopted by the architect. When the cladding failed, the client and the architect blamed the engineer. It was clear that the fixing system had not formed part of the engineer's retainer at the outset, but he had unwittingly extended his brief by his attempt to assist. He should have either refused to become involved or insisted that the architect accepted full responsibility for the specification.

Budgets and cost reductions

When tenders were received they were substantially above the figures predicted by the design team. The client could not afford to go ahead without first achieving reductions. The design team was asked to look for cost savings. They did so but the client wished to achieve a prestigious development and so would not compromise on finishes. Pressure was put on the team to produce savings which would not affect the appearance of the building. Changes were made to the basement structure and waterproofing system which later failed. A damp basement made it difficult for the client to find a tenant and so remedial costs and consequential loss of rent were suffered. The designers should have ensured that the client understood that the revised specification was of a lower standard than the original and the nature of the problems this might cause.

Quality control

An engineer who had difficulty recruiting sufficient experienced draftsmen and was under pressure because he was falling behind programme, started to send drawings to site without checking them or warning the client or contractor that they were unchecked. Inevitably mistakes had occurred because inexperienced staff were working under too much pressure. Delays to the project were caused and the contractor's claims were passed back to the engineer. Checking is essential and perhaps all the more so now that computer technology is so widely used, because inexperienced staff will not necessarily appreciate that the result produced by the computer may be wrong.

Inspection

The project was running late and everyone was under pressure to achieve the programmed completion date. The resident engineer was dissatisfied with the standard of work but agreed to accept the contractor's assurances that quality would be improved for the remainder of the work and did not require already completed works to be opened up. When the project was completed, problems were identified with concrete nibs designed to support brickwork cladding, and expensive remedial works were required. It was clear that the resident engineer had witnessed the bad workmanship but he had not acted upon it. He would doubtless have been criticised for failing to inspect sufficiently diligently if he had not in fact spotted the contractor's shortcomings. The only way to escape criticism as an inspector/supervisor is to make clear how often you will visit site and what proportion of the works you will see, to ensure that you do what you have promised to do, and that you follow up all incidents of bad workmanship. Inferior quality should only be accepted if the client has been warned of the possible consequences and has agreed to accept the risk.

Danger signs

Even when difficulties have arisen on a project, the right approach can prevent the situation getting out of control. An awareness of potential danger signs can help you respond appropriately. For example, look out for letters which seem to have been written with a view to being used as the basis for a claim later on. Guard against the breakdown of personal relationships, and if necessary introduce a new face to take the heat out of the situation. If your fees have been paid regularly and then payments stop, this may be because of dissatisfaction with your work. Query this immediately and try to resolve the underlying problem. This kind of situation should be reported to insurers who can provide guidance as to how to deal with it.

Equip yourself to deal with disputes which might arise in the future. Record decisions and the advice and any warnings given, and make sure such records are kept carefully, preferably for at least 6 years (12 years if your contract is in the form of a deed). Make sure that the minutes of meetings accurately reflect what was discussed and that correspondence critical of your performance is answered promptly.

When disputes arise

How to respond

However careful you are, some disputes will occur. It is a rare professional who has never had to make a call upon his professional indemnity insurance policy.

The construction industry is known for its adversarial attitudes and despite attempts by Latham and others to reduce the conflict, it is perhaps inevitable in an industry which deals with substantial sums of money and produces a unique product almost every time.

When you think there is a possibility of a claim, the first thing you should do is notify your insurers. Failure to do so can invalidate your policy. Remember that it is circumstances likely to give rise to a claim which must be notified, and not just the claim itself, so don't wait until legal proceedings are served upon you. In the light of the civil justice reforms of 1999, it pays to anticipate litigation and prepare for it by investigating fully as early as possible.

Methods of dispute resolution

Of course not all disputes are litigated. Far from it. The alternatives include arbitration, adjudication and alternative dispute resolution, and of course there is always scope for negotiation either before or during any more formal dispute resolution procedure.

Litigation

This is the ultimate means of resolving disputes and provides the element of compulsion necessary to ensure that the other party to the dispute engages in the process. Unfortunately, it has a reputation for being both slow and costly, although the new Civil Proceedings Rules introduced in 1999 were designed to improve this situation. The reforms were intended to make the process more cost effective and in particular to ensure that the costs are proportionate to the sum in dispute.

The majority of construction disputes of any size will be allocated to the Technology and Construction Court, which provides judges specialising in this field. The procedure is started with a claim form, followed by particulars of claim, served by the claimant. The defendant must then acknowledge service and serve a defence. Pre-trial protocols currently exist for personal injury claims, but there is a general requirement that parties observe the spirit of the protocols, whatever the nature of their dispute. This requires the claimant to give the defendant time to investigate the allegations made against him and to form a view on the merits of the claim and its financial value. The parties are intended to be much more open with one another than they were in the past, and it is increasingly difficult to sit on the fence and avoid engaging in a debate about the issues by refusing to explain the case fully, or by issuing bland denials. This may lead to earlier settlements, which save on legal costs and so is to be encouraged. If one party has a particularly weak case, it will bring forward the day of reckoning but will not necessarily lead to a worse financial outcome for them.

The parties must disclose to one another relevant documents, provided that the cost of doing so is proportionate, and the parties have to certify that they have made a reasonable search for such documents. Expert witnesses are commonly employed in construction disputes. There is a presumption in favour of single experts where appropriate, to avoid both the cost and the conflict which can arise when each party instructs its own expert. Where the parties each instruct their own expert, the experts will usually be required to meet with a view to agreeing facts and identifying areas of agreement and disagreement. Such meetings can be crucial in bringing home to the parties the strengths and weaknesses of their case and thereby assist in bringing about an out of court settlement which is one of the prime objectives of the new court rules.

Arbitration

Arbitration amounts to litigation in the private sector. It cannot be imposed upon an unwilling party, but consent will normally be given in advance by the incorporation of an arbitration clause in the contract, in which case any litigation can be stayed while the dispute is referred to arbitration. An arbitrator can be chosen by the parties, being nominated in the contract or selected once a dispute has arisen, or nominated by a nominating body, such as the Chartered Institute of Arbitrators.

The procedure is regulated by the Arbitration Act 1996, but is basically similar to the litigation process. The arbitrator must act fairly and impartially, but depending upon the terms of the arbitration agreement, he may have greater flexibility than a judge, both in relation to the evidence he can receive and the manner in which it is presented to him. He can take the initiative in ascertaining both the facts and the law.

It is usual to have a preliminary meeting to define the issues and discuss the procedure, but the hearing can be relatively informal. Awards can be challenged only on limited grounds, including lack of jurisdiction or if the decision is obviously wrong and of general public importance.

Arbitration can be difficult to arrange in any dispute involving more than two parties.

Alternative dispute resolution (ADR)

There is a variety of forms of alternative dispute resolution, of which mediation is the most common. Other possibilities include expert determination – involving the appointment of an expert to decide the dispute. If the parties so agree, the expert's decision can be binding. Dispute review boards are established under certain forms of building contract. They are usually set up for the duration of the project to determine all disputes arising during that period. You should, of course, discuss the situation with your insurers before agreeing that any form of ADR should be binding.

Mediation is consensual in that it is a form of structured negotiation conducted with the assistance of an independent go-between. It is non-binding and the parties can abandon the attempt to mediate at any point. It only becomes binding when a formal agreement recording the terms of settlement is signed at the conclusion of the mediation.

It can be both quicker and cheaper than other forms of dispute resolution, but preparation is just as vital as it is in litigation or arbitration, so the cost should not be regarded as negligible. Mediation is rapidly gaining in popularity, not least because the new court rules encourage it.

The normal procedure is for the parties to exchange introductory statements in writing, to meet together on the appointed day to restate their cases briefly and then retire to individual rooms while the mediator goes between them. What is said to the mediator is confidential, but he should encourage the parties to consider the risks which they will run if the mediation is not successful and the options available to narrow the gap between the parties. When this process results in an agreement, it will normally be put in writing and signed by the parties before they leave the mediation. At that point the deal becomes binding.

Adjudication

Certain building contracts have allowed for adjudication for some years, but since May 1998 when the Housing Grants Construction and Regeneration Act 1996 came into force, all construction contracts are required to contain an adjudication clause.

Construction contracts as defined in the Act include professionals' terms of appointment.

Either party can give notice of an intention to refer a dispute to adjudication at any time, and the dispute must then be referred to an adjudicator within seven days. The adjudicator has 28 days to reach a decision, unless a longer period is agreed by the parties. The decision is binding until the dispute is finally resolved by litigation or arbitration. The courts have shown a readiness to enforce adjudicators' decisions.

An adjudicator can be nominated in advance or a nominating body can be selected. The procedure during the adjudication is in the hands of the adjudicator, provided that he acts impartially. He can decide the procedure and take the initiative in ascertaining the facts or the law. Adjudication is ideal for disputes arising during a project which might otherwise hold up the progress of the works, and for fee recovery, but it is questionable whether the more complicated construction disputes can be adequately resolved in the 28 days allowed.

Preparation

Adequate preparation is a requirement common to all forms of dispute resolution. As mentioned above, preparation can start even before there is any suspicion that a claim will be made, by keeping proper records of discussions, meetings, inspections, etc. As soon as you are aware of circumstances likely to give rise to a claim, you should notify your insurers. At that stage it is not too early to start putting together the documentation which will help you to defend a claim if it is made, and if you do this you will also be prepared to put your point of view across to the other party to the dispute in any discussions or negotiations which take place before a formal claim is made. In this way you may even dissuade them from bringing a claim at all.

Once a formal method of dispute resolution has commenced, the information you have put together will be invaluable to your defence team. If a dispute is to be referred to adjudication, they will have very little time to get to grips with the issues and advance preparation will pay for itself many times over.

Apart from the documentation, potential witnesses need to be identified. Witnesses of fact must refresh their memories from the project files, or if the facts are still fresh in their minds should be encouraged to write them down so that important details are not forgotten. Provided this is done in contemplation of litigation, the resulting documents should not be disclosable, but it is perhaps better if you limit yourself to the points you would wish to make in your own defence, rather than producing anything in the nature of a confession.

Your insurers will take a prominent role in the resolution of any dispute. Leaving aside the excess and any uninsured sums, it is the insurers who will have to pay any compensation due and they will normally be responsible for the defence costs. For that reason, you should not attempt to negotiate a settlement without their prior approval. They will take an active interest, and may dictate the choice of lawyers and expert witnesses, but reputable insurers will bear in mind that commercial relationships, professional reputations and substantial excesses may be at stake and will wish to work in partnership with you in order to resolve the dispute. It is in fact unusual for there to be any significant disagreement between insurers and insured about how a dispute should be dealt with.

Conclusion

The claims experience is unlikely to be a pleasant one – nobody likes having their professional competence challenged. It can take up a great deal of time which would be more profitably spent developing your practice, but the process need not be entirely negative. Lessons can be learned from claims, whether yours or someone else's. That is part of the purpose of this book. Lessons can be learned if you are prepared to take an unbiased view and accept that you could have performed better, and your colleagues will learn from your mistakes if you do not hide from them the fact that mistakes were made.

8 Risk management from the lawyer's point of view - comments on US practice

Robert A. Rubin and Dana Wordes

Introduction

This brief chapter highlights the differences between American and British practice as applied to risk management for design professionals.

The previous chapter addressed risk management from a British lawyer's perspective. It discussed the impact of statutes such as the Housing Grants Construction and Regeneration Act and Contract Rights of Third Parties Act on the resolution of construction disputes. It referred to standard terms in commonly used industry contracts such as JCT 80 and the Association of Consulting Engineers (ACE) Conditions.

Although the American and British legal systems share the same roots, the United States does not have a national system of laws that apply to construction contracts. Federal regulations govern public contracts for federal agencies. State laws apply, in varying extent, to public and private contracts in the 50 states. In addition, courts in different states have different interpretations of issues that affect risk management for design professionals, such as the ability of third parties to sue for economic loss or the enforceability of contractual limitation of liability clauses.

In the United States, form contracts such as those developed by the American Institute of Architects (AIA) or Engineers Joint Contracts Documents Committee (EJCDC) are frequently used. The AIA and EJCDC documents are not industry standards and they are extensively modified for most contracts. Large public and private owners often use their own standard contracts. On large, unusual or high-profile projects, the parties may prefer to negotiate manuscript agreements.

We found differences between British and American practice in the areas of insurance, limitations of liability, rights of third parties, termination of the contract and dispute resolution. The American contract terms relating to the design professional's responsibilities were generally equivalent to the British terms. The chapter includes typical American contract clauses that are similar to comparable British clauses.

Insurance and indemnity

Contractors

In the United States, professional indemnity insurance for contractors is not readily available. In recent years, insurers have responded to industry pressures to provide insurance coverage for design and build projects, and projects where contractors assume responsibility for design of structural steel connections, curtain walls and fire suppression systems.

AIA A-201-1997 (owner-contractor agreement) paragraph 11.1.1 lists the items that must be included in the contractor's liability insurance, such as coverage for workers' compensation and benefits, motor vehicle, personal injury and property damage insurance, and insurance for liability assumed in the contractual indemnity clause.

AIA A-201-1997 paragraph 3.18 is a typical indemnity clause:

To the fullest extent permitted by law and to the extent claims…are not covered by…insurance…purchased by the Contractor…the Contractor shall indemnify and hold harmless the Owner, Architect,…and agents…from and against claims, damages, losses and expenses, including but not limited to attorney's fees, arising out of or resulting from performance of the Work, provided that such claim, damage, loss or expenses is attributable to bodily injury, sickness, disease or death, or to injury to or destruction of tangible property (other than the Work itself) but only to the extent caused by the negligent acts or omissions of the Contractor.

Design professionals

EJCDC 1910-1[*] (owner-engineer agreement) paragraph 6.05 requires the engineer to buy and maintain insurance. The EJCDC form lists various types of insurance that the owner may require. Interestingly, the provisions listed on the form do not include professional liability insurance (EJCDC 1910-1 exhibit G). Sophisticated owners tend to require professional liability insurance.

The engineer must indemnify the owner for all losses 'caused solely by the negligent acts or omissions of the Engineer' or its employees or consultants (EJCDC 1910-1 paragraph 6.11.1). However, where the engineer is not solely at fault, the engineer's liability is proportional to its negligence (EJCDC 1910-1 paragraph 6.11.3).

Latent defect insurance

The authors are not aware of latent defect insurance. However, latent defects may be covered under other insurance policies, such as completed operations coverage, provided that all policy conditions are met.

Professional liability policies are generally issued on a claims made basis. Claims are covered if they are asserted and reported during the policy period and arise out of

---

[*] 1996 edition referred to unless specified otherwise.

professional services performed after the retroactive date. The retroactive date usually includes the date of the first policy with the current carrier; it may not provide coverage for periods covered by previous carriers.

Project insurance generally runs for the life of the project and a specified period after substantial completion of construction. It covers the vicarious liability for construction operations for all project team members, but typically excludes coverage for professional services.

Limitations of liability

Statute of limitations and statute of repose

Every jurisdiction in the United States (court systems in each of the 50 states and the federal courts for federal contracts) has its own statutes of limitations. These statutes limit the time in which suit must be brought after the cause of action accrues. The 'cause of action accrues' when the triggering event gives a party the right to sue. Different states have defined accrual differently. However, for suits by parties to a construction contract, the trigger is usually substantial completion of construction or completion of the design professional's services. For suits by non-parties, the trigger may be the date of injury or, in some states, when a reasonable person should have discovered the injury, the 'discovery rule.' The statute of limitations varies for different types of claim (e.g., contract, tort). Contractual time limits that are shorter than the statutory periods are generally enforceable.

Most states have enacted statutes of repose. These differ from statutes of limitations in that the time limits are not measured from the time of injury. Instead, statutes of repose set an absolute time limit for bringing suit, generally a specified number of years after substantial completion, regardless of when the injury occurred. For example, if the statute of repose was 10 years and building collapsed 20 years after construction, an injured bystander could not recover from the designer or contractor based upon negligent design or defective construction.

Contractual limitation of liability

In the United States, contractual limits of liability are generally enforceable. The EJCDC offers three alternative limitation of liability clauses. An engineer may limit the liability to the owner to the amount of the engineer's total compensation. Alternatively, the liability may be limited to the amount of contractually required insurance, or the total liability may be limited to a specified amount (EJCDC 1910-1 exhibit 1, paragraph 1).

In addition, the EJCDC provides sample language to limit the types of damages by a general exclusion of consequential damages or by specific exclusions such as the cost of replacement power or loss of revenue (EJCDC 1910-1 exhibit 1, paragraph 2). The AIA contracts contain a blanket mutual waiver of consequential damages (AIA B-141-1997 paragraph 1.3.6).

Rights of third parties

Under the common law, a third party may enforce a contract if it is an intended beneficiary of the contract. A third party would be an intended beneficiary if both parties to a contract intended that it be able to enforce the contract and if performance of the contract was owed directly to the third party. Potential beneficiaries to construction contracts include
  • contractors that allege that the delays due to the architect, engineer or another contractor impacted upon their operations
  • subcontractors that seek to enforce the owner's contract with the contractor, and
  • end users of a project, such as condominium purchasers, that seek to enforce the owner's contracts with designers or builders.
Generally, courts find that these third parties are incidental beneficiaries of the contract and are unable to enforce it.

The authors are not aware of an American counterpart to the Contracts (Third Party Rights) Act, which gives third parties with an interest in a construction contract the right to enforce the contract if the contract expressly provides it or purports to confer a benefit on the third party.

Most American contracts expressly limit the rights of third parties. EJCDC 1910-1, paragraph 6.08.C. indicates that, unless otherwise expressly provided for, 'nothing in this Agreement shall be construed to create, impose, or give rise to any duty owed by the Owner or Engineer to any Contractor, Contractor's supplier, other individual or entity, or to any surety for or employee of any of them.' Furthermore, duties undertaken under the agreement are for the 'sole and exclusive benefit of Owner and Engineer.' Similarly, AIA B-141-1997, paragraph 1.3.7.5 states that 'nothing contained in this Agreement shall create a contractual relationship with or a cause of action in favor of a third party against either the Owner or the Architect.'

Even where the contract is silent with respect to the rights of third parties, third parties may be unable to recover purely economic losses. Many state courts have adopted some form of the economic loss rule, which generally bars third parties to a contract from recovering in tort for purely monetary loss. Some states make a limited exception if the loss is due to the design professional's negligent misrepresentation and the relationship is the functional equivalent of privity of contract. Other states permit recovery by third parties whose pecuniary losses are foreseeable.

Termination

EJCDC 1910-1 paragraph 6.06 permits either party to the contract to terminate for cause upon 30 days written notice. In addition, the engineer may terminate upon seven days written notice if the owner requests it to 'furnish or perform services that are contrary to the Engineer's responsibilities as a licensed professional.'

AIA B-141-1997 paragraph 1.3.8.4 permits either party to terminate for cause upon not less than seven days written notice if either party fails to substantially perform.

Professional responsibilities

Form contracts prepared by professional groups such as the AIA and EJCDC integrate the design professional's responsibilities under its contract with the responsibilities indicated in the owner-contractor agreement.

EJCDC paragraph 6.01 states that the engineer will use 'the care and skill ordinarily used by members of Engineer's profession practicing under similar circumstances at the same time and in the same locality.' It indicates that the engineer makes no warranties, express or implied.

In contrast, AIA B141-1997 (owner-architect agreement) makes no representations about the architect's duty of care. It does, however, state (paragraph 1.2.3.2) that the 'Architect's services shall be performed as expeditiously as is consistent with professional skill and care and the orderly progress of the Project.'

Supervision of construction

Engineers and architects generally seek to limit their responsibility for construction. EJCDC 1910-1 exhibit A, paragraph A1.05 (A) (6) indicates that the engineer shall visit the site at intervals appropriate to the stages of construction. The purpose of the engineer's site visits is to 'enable the Engineer to better carry out the duties and responsibilities assigned to and undertaken by Engineer…and…to provide…a greater degree of confidence that the completed Work will conform in general to the Contract Documents.' Further, the engineer shall not have authority over or responsibility for the contracor's means and methods of construction or site safety.

Similarly, AIA B141-1997 paragraph 2.6.2.1 indicates that the architect shall 'visit the site at intervals appropriate to the stage of the Contractor's operations…to become generally familiar with and keep the Owner informed about the progress and quality of the…Work…However, the Architect shall not be required to make exhaustive or continuous on-site inspections to check the quality or quantity of the Work.' In addition, the architect shall not have control over or responsibility for the contractor's means and methods of site safety. Paragraph 4.2.2 of AIA A201-1997, the owner-contractor agreement, repeats this language.

Dispute resolution

Adjudication

The United States has no counterpart to the British requirement for adjudication of construction disputes. However, many public agencies have their own expedited dispute resolution procedures. These procedures usually cast the design professional or contractor as the claimant. The claimant generally has an obligation to produce documents that support its claim, without a corresponding requirement for the authority to produce its documents. Disputes are often decided by a member of the agency and the scope of judicial review is limited.

Contractual dispute resolution procedures

The AIA contracts mandate mediation as a condition precedent to mandatory arbitration (AIA B-141-1997, paragraphs 13.4, 13.5; AIA A-201-1997, paragraphs 4.5–4.6). In contrast, the EJCDC contracts require 'good faith negotiations' for 30 days before instituting legal or contractual dispute resolution proceedings (EJCDC 1910-1 paragraph 6.09). The suggested procedures for contractual dispute resolution include mediation (as a condition precedent to litigation) or arbitration (without mediation) (EJCDC 1910-1 exhibit H).

Arbitration

The federal government and most states have arbitration acts. These acts generally address the enforceability of a demand for arbitration, enforceability of an arbitration award, the basis for judicial review of arbitration awards and the limited grounds for vacating an award. The statutes do not address arbitration procedures. In the construction industry, arbitration procedures are generally governed by the American Arbitration Association's (AAA) Construction Industry Arbitration Rules because the rules are incorporated by reference.

Procedures for arbitration

The AAA's construction ADR task force developed a three-tier structure of arbitration procedures comprising the standard or regular track, and two variations. Under the AAA's regular track procedures, the party initiating the arbitration (claimant) starts the process by
1. giving written notice to the other party (respondent) and
2. filing the notice with the AAA.
The AAA becomes the arbitration administrator. The AAA then officially notifies the respondent, who must file an answer and counterclaims with the AAA and the claimant within 10 days. A claim or counterclaim may be amended any time before the appointment of the arbitrator. Arbitrators are appointed by a court or administrative body such as the AAA or selected by the parties. The parties may choose candidates by eliminating unacceptable names from a roster of neutrals. Another method of selection requires each party to select an advocate arbitrator. The advocate arbitrators in turn elect a neutral arbitrator to complete their three-member panel.

After filing the preliminary claims and counterclaims, disputants may participate in administrative conferences and preliminary hearings to expedite the actual arbitration proceedings. As in litigation, many disputes are settled before formal hearings are held.

The arbitration may proceed as an open hearing or through written transmission of evidence and position papers to the arbitrator(s). The determination of an arbitrator results in a binding award. According to AAA rules, an award will be announced no later than 30days after the close of hearings.

The AAA has also promulgated 'fast track' procedures to expedite smaller and less complex cases. The fast track is intended for multiparty cases when no party's claim or counterclaim exceeds $50,000 (excluding costs and interest) or by agreement of the parties. However, fast track procedures are available for all two-party disputes regardless of the claim amount.

Under fast track, the arbitrator appointment process is streamlined by limiting the time for disputants to object to an AAA roster candidate. In addition, the hearing is scheduled within 30 days of the arbitrator's appointment, and the award is rendered no later than seven days after the hearing. Fast track limits extensions of time to file and amend claims or counterclaims and limits discovery to an exchange of exhibits before the hearing.

In order to provide more sophisticated procedures and expertise for the growing caseload of large multi-party arbitrations, the AAA developed the large, complex case program (LCCP). If no party objects, cases in which any claim or counterclaim exceeds $1 million (excluding costs and interest) are administered under the LCCP. The main features of this process are expanded management provisions and a roster of elite neutrals trained specifically in the resolution of large complex cases. As with all arbitration, the rules of an LCCP proceeding are subject to the arbitrator's discretion. However, the structure allows for greater latitude on discovery and scheduling.

Dispute review boards

Dispute review boards (DRB) are widely used for underground construction and where the cost of construction exceeds $100 million. They are less common in building construction projects. A DRB is an integral part of a construction project. The requirement for a DRB is included in the contract documents, and the owner, contractor and members of a DRB sign a three-party agreement obliging them to use the DRB to attempt to resolve disputes while construction is ongoing. The board consists of three members: one each chosen by the contractor and the owner and approved by their counterpart and the third, who usually serves as the board chairperson, is selected by the other two and approved by both. The board receives periodic progress reports and may occasionally visit the site to get first-hand knowledge of the progress of the work and potential disputes. The owner or contractor may request the board to participate in the resolution of a dispute. The board may review contract documents, hold hearings with project personnel and make a non-binding recommendation. In contrast with mediation proceedings, the DRB proceedings are typically admissable in court in the event that the parties do not accept the Board's recommendation.

Expert witness practice

American practice usually is for each party to retain its own experts. Courts rarely appoint independent experts. During the pre-trial discovery phase, each party furnishes information about the credentials of the experts who will be used at trial, the general scope of their testimony and the basis of their testimony. In contrast to new rules described in the previous chapter, opposing experts in the United States rarely discuss technical issues with each other, unless such meetings are part of settlement negotiations.

However, an expert may be subpoenaed to testify at a pre-trial deposition. Opposing counsel (and its experts) will assess the expert's credibility and demeanour, as well as the strength of the expert's technical position. Often, the assessment of each party's experts will influence whether the dispute will settle before trial.

In construction litigation, the expert witness performs two functions. First, the expert testifies to his expert opinion. Second, and equally important, the expert educates the court or jury about technical issues to enable them to understand the expert opinion. Education is a critical step in making the expert's opinion valuable to the trier of fact. When one party to litigation calls an expert witness, the opponent usually calls his own expert. The opposing party sometimes calls the expert witness with the hope of confusing and obscuring the issues to such an extent that the trier of fact will disregard the expert evidence.

Experts may participate in ADR proceedings. In arbitration, the forensic engineer's testimony would be similar to expert testimony given at a trial. In mediation, the expert's presentation may be less formal. The expert may explain technical issues and emphasize how strong the case would be if the parties were unable to settle and the expert testifies at trial.

9 The technical investigation of failure – a marine industry perspective

J.S.Carlton and J.R.Maguire

Introduction

The Technical Investigation Department (TID) of Lloyd's Register (LR), since its formation 50 years ago, has been invited to investigate some of the more difficult engineering problems posed by the marine and other industries that Lloyd's Register serves. More recently, this subject area of engineering failure investigation has become fashionably known as forensic engineering.

Within TID, investigations have been accomplished by a combination of trial measurement, metallurgical investigation, theoretical analysis and engineering judgement based on the department's accumulated experience of over 5000 investigations. Table 9.1 shows an analysis of the distribution of investigation types that have been undertaken in each of the marine, land-based and offshore industry sectors. This experience has embraced a wide range of failure and mal-performance based engineering problems. Furthermore, the TID archives bear adequate testament to the cyclic nature of failure problems. Indeed, in TID's experience, it is a general characteristic that investigations tend to be grouped into particular technical subject areas for discrete time intervals which then re-occur periodically.

To address these problems TID comprises a multidisciplinary group of engineers and scientists which includes mechanical, marine, civil and structural engineers, naval architects, physicists, electrical and electronic engineers, mathematicians, computer scientists and statisticians. The talents of this relatively unique group of highly qualified engineers and scientists are combined to provide a high quality, rapid response, technical consultancy and advisory service to the marine industry, the other industries that LR serves and to LR itself. These activities are supported by fully equipped instrumentation and material investigation laboratories.

The department also conducts medium and long-term marine engineering research and development to equip LR, in keeping with its divisional strategies, with the necessary capabilities to meet future requirements based on a continuing review of technological development and achievement. Clearly, by the nature of the department's primary work, many of the research initiatives derive directly from the failure scenarios presented to TID.

Table 9.1 *Primary classification of investigations*

Marine	%	Land-based industries	%	Offshore	%
Hull/machinery vibration	25.8	Petrochemical /industrial	21.7	Platforms	58.3
Shafting	24.1	Power stations	20.3	Miscellaneous	33.3
Diesel engines	14.1	Miscellaneous	18.5	Pipelines	8.4
Gearing	9.1	Buildings	17.0		
Power absorption	5.8	Nuclear	14.5		
Noise	2.7	Docks and harbours	4.7		
Turbines	2.5	Bridges	3.3		
Electrical	2.4				
Propellers	2.1				
Auxiliary machinery	2.0				
Condition monitoring	1.9				
Boilers	1.6				
Couplings/clutches	1.5				
Ship structural failure	1.2				
Rudders	1.1				
Environment	0.9				
Miscellaneous	0.8				
Pipelines	0.4				

Philosophy of investigation

The diagnosis of engineering failure is an art that has to be acquired over many years by investigation engineers; indeed, it is a process that is honed throughout an engineer's working life. As such, the most senior and experienced investigation engineers are those who have devoted considerable portions of their lives to this area of technology. This, however, is not to devalue the contributions of engineers and scientists who have spent lesser amounts of time in this field of endeavour before moving on into the other branches of technology with a considerably enhanced experience.

In their essentials, however, the principles of investigations are no different from the fundamental diagnosis techniques learnt by medical practitioners. The prerequisite for diagnosis is to enter an investigation with an open mind, no matter how many apparently similar exercises have been undertaken in the past. Minor changes in environmental circumstances, loading, manufacturing or design detail can bring entirely new perspectives and subtle twists to an otherwise familiar failure scene. Nevertheless, initial impressions, indications and signs, provided that they are ultimately rigorously challenged in the hypothesis testing procedure, can be extremely valuable in the initial stages of an investigation.

Most diagnosis situations start with the formulation of a hypothesis about the cause of a failure against which the various facts of the case are eventually tested. If one fact is found not to fit, then either the hypothesis is rejected or modified. As such, failure diagnosis is an iterative process. The initial working hypothesis is formulated from various sources:

- Accounts of the failure by the owners of the problem and those close to it. This clearly is essential evidence. However, it is usually also filtered evidence in which the features considered to be important by the reporter are presented or accentuated. Clearly, each of the symptoms highlighted needs to be probed in order to reveal its true significance or, indeed, reveal further symptoms and avenues of exploration. Considerable skill is required at this stage of the investigation on the part of the investigator if the correct path or paths are to be followed at the start of the investigation.

- A viewing of the failure evidence is essential either at first hand, which is the most preferable, or by photographs and logged accounts where the evidence is inaccessible. The examination of failure evidence needs to be undertaken in the most thorough way possible if crucial evidence is not to be lost. In many cases a 're-construction' of the failed parts can prove a most helpful and revealing exercise. Photographic evidence is frequently difficult to assess effectively since photographers are, by their training, often most concerned with producing pictures having considerable artistic rather than scientific merit. Equally, images of submerged failures are frequently incomplete and unclear due to the constraints under which they are obtained. However, it should be recognised that considerable progress in this area has been made in recent years.

- A consideration of the engineering and environmental principles likely to be applicable to the problem in terms of, for example, the ambient conditions, the corrosiveness or otherwise of the environment, the likely static and dynamic load paths through the item under consideration. By way of illustration, in this latter context, a most useful basic technique in a ship vibration investigation, prior to undertaking an in-depth exploration of the problem, is to *walk the ship* using the soles of one's feet or a very simple portable instrument in order to understand the various global characteristics of the problem before considering the detail. The importance in an investigation of global understanding of the engineering interactions is considerable prior to concentrating on the detailed aspects.

- The service records and how these relate to similar types of components or systems. In particular, whether there has been a previous history of failures or indications of a growing problem in the time leading up to the failure being investigated.

- The investigator's experience of previous case histories.

From an initial working hypothesis, a suitable instrumentation fit and trials programme may be prepared; sometimes in the form of a pilot or exploratory investigation. Alternatively, a theoretical model may be derived to test the validity of the hypothesis. In either case, if it is a failure investigation, a metallurgical investigation of the failed component usually proceeds in parallel with information gained from the other various lines of enquiry as they become available. The metallurgical investigation aims at establishing the material chemical composition, structure, failure mode and the position of the origin and the presence of foreign bodies, voids, corrosion products and so on. In this respect keeping the failed specimens free from contamination after the failure event is of the utmost importance, even though the specimens may be of considerable size.

In the case of an investigation involving a measurement programme, following the full scale trials stages of the investigation, an analytical or numerical model may be formulated in order first to obtain correlation with the quantitative trials data set and then, secondly to extend this domain so as to formulate proposals for appropriate remedial action or gain a deeper understanding of the failure processes. Typical of such a situation may be where measurements are constrained for some operational reason to be taken in, say, a ballast condition for a ship fitted with a fixed pitch propeller and data are, therefore, unavailable for the deep loaded conditions.

Alternatively, analytical models are frequently used to complement measured trials data in terms of, for example, general validity and identifying mode shapes of vibration. For theoretically biased investigations, a similar procedure applies but without the benefit of measured data, the correlation being undertaken by comparing the predictions with the observed qualitative behaviour and other known standard test cases. In these cases significant demands are placed on the selected analytical or numerical procedures which, in turn, require that the chosen methods are well correlated and that their predictions can be generally supported by less detailed methods or heuristic insights into the appropriate behaviour. The essential feature, however, of any computational method, whether used in association with field trial measurements or in isolation, is that it has been properly validated and correlated with the results of model or full scale experience.

By way of example of the divergence that can occur between the results of computations obtained from reputable organisations, Fig. 9.1 shows the propeller blade stresses calculated using finite element methods for a highly skewed propeller blade. The experimental values were obtained from work carried out in TID using a 254 mm model of a 72 degree biased skew propeller under the action of point loading. An alternative example of reasonable correlation is seen in Fig. 9.2 which shows a comparison between the results of finite element and boundary element computations and the results of three-dimensional photo-elastic experiments. This exercise formed a precursor to an investigation on the integrity of some nuclear pressure vessels.

For any of the investigation scenarios described above the accumulated experience of past failures, and the factors involved in their manifestation, are an essential ingredient. Using these techniques a number of iterations are performed until all of the known facts of the case can be supported from the evidence gathered or derived. Consequently, a unique blend of field measurement, metallurgical investigation, theoretical analysis and engineering judgement combined with a sound historical knowledge of failure situations are required to satisfactorily execute an investigation. Within this process, however, the more complex measurement tasks are engineering projects in their own right, requiring design, manufacturing and installation phases which must fit with other technical requirements and production or repair schedules.

When undertaking investigations of failures it is important to recognise that the failure may not be attributable to a single cause. Frequently, two or more contributory factors may be involved, the combined effects of which have resulted in producing the conditions for failure. Therefore, the approach taken to an investigation must always allow for this possibility which in practice frequently occurs. In this context the investi-

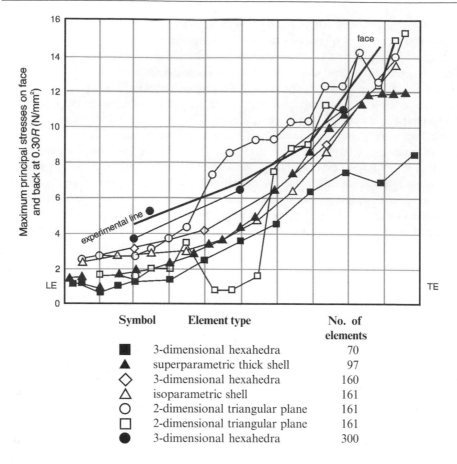

Symbol	Element type	No. of elements
■	3-dimensional hexahedra	70
▲	superparametric thick shell	97
◇	3-dimensional hexahedra	160
△	isoparametric shell	161
○	2-dimensional triangular plane	161
□	2-dimensional triangular plane	161
●	3-dimensional hexahedra	300

Figure 9.1 *Comparison of different propeller blade stress calculations with experimental data.*

gator must always endeavour to understand the underlying causes of the problem because if this is not done, then proper judgement cannot be exercised on whether palliative or fundamental changes to the system are required.

Two case studies

In order to illustrate the diversity of investigations the following two studies are cited. The first outlines a relatively straightforward investigation, although not trivial, of a diesel engine failure and involves a reconstruction of events. The second involves two apparently similar cases but requiring major instrumentation exercises, extensive field trials, theoretical studies, metallurgical investigation and model tests in order to understand the problems involved and ultimately leading to differing causes of the failures experienced.

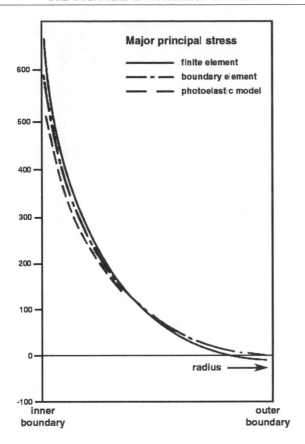

Figure 9.2 *Comparison of finite element, boundary element and photo-elastic experiments for a nuclear vessel.*

Loose crank bearing bolt

TID surveyors are often called upon to attend the site of a mechanical mishap to conduct an investigation into the cause. Sometimes the cause is simple, as in this case of a loose crank bearing bolt.

The engine which suffered damage was an eight cylinder two-stroke diesel engine arranged to drive a large alternator. The first indications that it had a problem were loud knocking sounds during normal operation. Unfortunately, before the engine could be brought to a stop, one of its connecting rods had abandoned its relationship with the crank pin and emerged from the crankcase, causing substantial internal and external damage. The final state of the connecting rod can be judged from Fig. 9.3.

As the various damaged parts became accessible from the wreckage, possible alternative causes were continually tested for credibility. By this process, attention was

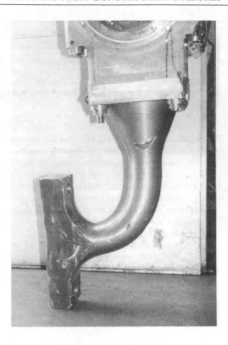

Figure 9.3 *Damaged connecting rod.*

increasingly drawn to the need to find the crank bearing bolts, which were not initially accessible. When they were retrieved, one of them was found to be virtually intact, but without its nut. This nut, its locking device and locking screw were all found separately. The nut threads for about one and a half turns nearest to the lower face were seen to have been damaged by axial shear. About one quarter of the first turn, although present, had been separated from the nut around its circumference. The other bolt had broken under severe combined tensile and bending load and its nut with locking device and screw were still present on the broken end. Furthermore, a bolt in a different connecting rod was found to be 1 mm slack with its nut locking device and locking screws correctly fitted.

This evidence clearly showed that the slackening and loss of the intact bolt were the initiating cause of the engine damage. In searching for the reason why the bolt became slack, the effects on the bolt of possible cylinder malfunctions were calculated and physical tests of similar bolts were conducted. The key finding from this work was that if the bolt had been properly tightened, the probability that sufficient torque could have been generated to loosen the nut against the frictional resistance of the threads and of the nut face was minimal. The torque due to tensile load, which normally acts to unscrew a nut, was found to be one eighth of the frictional torque resisting such movement. It was concluded, therefore, that the damage resulted because the bolt had not been properly tightened to the correct tensile load.

The routine maintenance procedures adopted for this installation involved periodic checks of the bolt tension. The method used was to slacken and re-tension the bolts and, as such, this procedure had the potential to introduce error.

Sterntube bearing failures in two passenger ships

These two investigations, which were undertaken within months of each other, illustrate how two apparently similar sets of initial symptoms led to conclusions which, although having much in common, had important differences.

In the first case, at the time of its sea trials the ship experienced, over a period of three to four months, a series of A-bracket bearing failures through overheating. After each failure, modifications to the propulsion line shafting were carried out by the builder which included re-alignment of the shafting, changes to bearing clearances and materials, and so on. Additionally, during the trials following each bearing modification, using data measured from the ship's own standard instrumentation, certain basic correlations between, for example, course changes and bearing overheating were deduced. However, after the fourth bearing failure a full investigation of the problem was commissioned from TID by the shipbuilders.

The investigation commenced with a theoretical analysis, using a vortex lattice model of the propeller operating in a scaled model wake field of the ship. These calculations indicated that the effects of the forces and moments generated by the propeller on the shaft dynamic alignment were likely to be significant in the straight ahead condition. Simultaneously with this computational study, a series of model tests were commissioned to explore the behaviour of the in-plane wake field components at the propeller location when the ship was turning. The data derived in this way were used both to gain an understanding of the importance of the cross flow components during turning and for subsequent theoretical computations for eventual correlation with the results of a full scale measurement exercise, which was also initiated at this time, and directed towards quantifying the shaft line dynamic behaviour. Both of these major experimental tasks within the overall investigation programme were required to be complete within a time frame of around four weeks.

The full scale measurement task required complex instrumentation to be installed on the ship (Fig. 9.4). This, in turn, needed custom built instrument housings and fixings to be fabricated and installed, underwater and conventional air-gap telemetry systems to be fitted together with a range of other less complex instruments and a suitably responsive data recording system with adequate storage. The ship was dry-docked to allow fitting some of these instruments, principally those to be used outboard of the hull and shaft sealing arrangements. Furthermore, as a part of the investigation the tail shafts were removed from the ship, transported 10 km and checked for concentricity in a suitably sized lathe. During this dry-docking period the stiffness of the ship's A-bracket arrangement was also determined by transversely recording displacements while pulling the bossing of the A-bracket from the dry-dock wall by pneumatically-operated chain blocks with a 100 tonne load cell coupled into the system. This sub-component of the trials programme required a civil engineering appraisal of the capability of the ageing dry-dock wall to withstand this treatment. Additionally, a series of viewing windows were cut into the hull above the propellers in order to observe the propeller behaviour, chiefly through its cavitation performance and with the aid of tufts on the hull and appendages, to see if any abnormal flow regimes developed. As such, all of these activities together with their attendant logistical problems of the transport of personnel and equipment over several thousand miles and the customs and importation formalities required to be phased in during the time allotted for the investigation so that not only the planned delivery of the ship could take place, but that delays to other ships wishing to use the dry-dock facilities would not be incurred.

Following the sea trials the detailed analysis of the relevant data gained took place. At the time of collection the investigation team undertook a preliminary analysis of all the data and the correlation with the theoretical predictions of the hydrodynamic and shaftline dynamic models. This involved the direct comparison of the data sets involved and the extension and interpretation of the measured data together with the positive identification of shaft vibration mode shapes. The principal findings of the investigation relating to the first ship were that the slopes of the shafts through the bearings were too high due to the dynamic loads generated by the propellers both in the normal ahead and turning conditions. Additionally, a blade order shaft lateral resonance was moved into the running speed range on the inside shaft when turning, due to the change of bearing loads induced by the hydrodynamic loading of the propellers. Of particular importance in this investigation was quantification of the entirely different characteristics of the loading vectors on each shaftline when the ship was turning, their

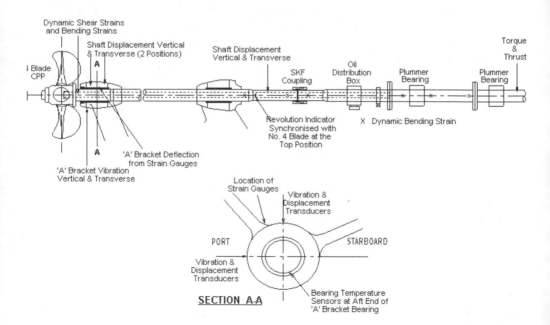

Figure 9.4 *Shaft instrumentation for the first ship.*

correlation with the observed burning marks on the bearings and the implications for this and future designs.

The second of the two ships cited exhibited very similar initial symptoms to the first and indeed, with the exception of the findings relating to the lateral resonance, the principal conclusions from the first investigation applied in this case. The investigation for the second ship followed very much the same format as in the previous case but without some features such as the model tests, full scale flow visualisation and A-bracket stiffness determinations. One notable difference was that the second ship was fitted with water- rather than oil-lubricated bearings. During the dry-docking period it was noted that craze cracking of the shaft liner material had taken place in way of the bearing retaining rings implying that contact between the retaining rings and shaft had taken place, due to the rapid wear-down of the bearings. Due to the mix of materials involved this introduced a potential for a grain embrittlement mechanism to take place in the shaft liner. Of additional concern was that a metallurgical analysis of a relatively uniform distribution of metallic particles along the length of the A-bracket and stern tube bearing materials identified the particles as coming from the shaft liner material. The distribution of these particles could not, however, be entirely explained by considerations of the likely water flow paths through the bearings and, as such, their presence raised questions of material incompatibility. The actual mechanisms operating in this case are, as yet, not completely understood and are the subject of a continuing research programme within Lloyd's Register with the full cooperation of the other parties involved. As a pragmatic measure in order to return the ship to sea the bearing material was changed and the shaft liner dressed to remove the effects of the craze cracking and potential grain embrittlement. To date the ship appears to be working satisfactorily.

Common failure modes

The majority of failure investigations encountered today by TID in the marine industry involve aspects of fatigue. Other failure modes such as brittle failure and creep are less frequently encountered with pure ductile failures being a rare event.

Fatigue failures

Fatigue failures require cyclic stresses to be experienced by a discontinuity in the material. Furthermore, the fatigue life of the component will be influenced by the magnitude and nature, compressive or tensile, of any steady mean stresses. The material type, local geometry, mode of manufacture and the environment in which a component operates also affect the fatigue strength.

Fatigue failures are normally characterised by being transcrystalline and occur without significant plastic deformation. Moreover, the general appearance of a fatigue failure, as distinct from its orientation, will be similar for most modes of loading, for example, whether the failure has been induced by axial, bending or torsional load-

Figure 9.5 *Typical fatigue failure: typical propeller blade failure (left) failure of a diesel engine gudgeon pin (right).*

ing or a combination of these. Figure 9.5 shows two typical fatigue failures, one on a marine propeller blade and the other on a diesel engine gudgeon pin. Fatigue failures are normally considered to progress as three distinct phases in ductile materials.

Stage I is strongly influenced by the slip characteristics of the material, the applied stress level, the extent of crack tip plasticity and the characteristic microstructural dimensions of the material. Where cracks are initiated in ductile materials it is generally considered that the cracks grow cyclically by the deformation in the slip bands near the crack tip which leads to the creation of new crack surfaces by the mechanism of shear de-cohesion. In components with relatively small flaws it has been found that cracks can spend a considerable time in this mode of the development. Typically, for a propeller blade this stage might account for some 80 to 90% of the crack life.

Stage II is where the plastic zone at the crack tip extends over many grains due to the higher stress intensities. This is in contrast to the *Stage I* mechanism which embraces only a few grains at a time and is essentially a single shear type of mechanism in the direction of the primary slip system. For *Stage II* crack growth the process involves simultaneous or alternating flow along two slip systems in the material. This duplex slip mechanism results in a planar crack path which is normal to the far-field tensile strain direction and hence defines the orientation of the crack face. For components or structures where there are significant initial flaws the major part of the fatigue life is spent while the crack is growing in this mode. Consequently, for welded structures *Stage II* crack growth laws and codes are especially applicable. One frequent characteristic of *a Stage II* fatigue failure is the presence of striations in the fracture surface the spacing of which, within the applicability of the Paris Law regime of crack growth, has been shown to correlate with the measured average rate of crack growth per cycle. Figure 9.6 shows a set of striations relating to a fatigue failure in the tooth of a gear wheel of an ice breaker. Of particular interest is the relatively coarse nature of some of the striations in relation to the normally observed magnifications of around 1000 to 4000 for this type of marine component. However, it is important to note that striations are not always present and it has been shown that environmental effects can influence their development. In air, for example, striations are clearly seen in pure metals, some ductile alloys and in many engineering polymers, but this is not always the case in steels and they can often be indistinct in cold worked alloys. In a vacuum, striations are not seen in a number of alloys which would normally be expected to exhibit them in air. Furthermore, crack growth rates can be an order of magnitude slower.

The final phase of fatigue crack growth is *Stage III* and this is where the crack has grown to a sufficient size where the component can no longer withstand the mechanical loads imposed upon it and, therefore, it fails by another mode. This final phase in the fatigue crack life is normally very short, of the order of microseconds for most dynamic situations.

Striations should not be confused with beach marks which are often clearly visible on fatigue fracture surfaces (Fig. 9.5). Beach marks are normally associated with periods of crack arrest whereas striations can be thought of as being due to the plastic blunting of the crack trip upon each cyclic application of tensile stress.

The propensity for crack growth is influenced by the environment due to the deleterious effect of corrosion when combined with stress. If the stresses are sensibly constant, as for example in a pressure vessel, the corresponding failure mechanism is usually termed stress corrosion cracking whereas the term corrosion fatigue is usually applicable when cyclic stresses are predominant. In practice, such clear distinctions are unusual because most components experience some stress fluctuations and significant residual tensile stresses can be induced by manufacturing processes. All environments are corrosive to some degree, even air and pure water, and it is interesting to note that the fatigue strength of samples is increased if they are tested in a vacuum. The crucial function of the corrosive action is its contribution in overcoming the stronger microstructural barriers during the early stages of crack growth. Such action is a chemical function (fluid composition,

Figure 9.6   *Fatigue crack striations on a gear tooth failure (× 500).*

pH value and electropotential) and is consequently time dependent which helps explain why there is no conventional cyclic stress limit for corrosion fatigue conditions. Also, fatigue life is not related solely to the number of stress cycles and the results of corrosion fatigue tests are influenced by the frequency of the applied loads. Corrosive agents include both acidic and alkaline embrittlement media: chlorine, sodium and sulphur are commonly encountered in marine and industrial failures which are caused by corrosion assisted cracking.

Fatigue research has more recently concentrated on short cracks and it has been shown that the growth rates of small flaws can be significantly greater than those for long flaws when considered in terms of the same nominal driving force. Figure 9.7 shows in schematic form the typical fatigue crack growth behaviour of long and small cracks at constant values of imposed cyclic range and load ratio. In this context small flaws are considered in terms of one of four classifications (Suresh and Ritchie, 1984):

- *Microstructurally small:* those where the crack size is comparable to characteristic microstructural dimensions.
- *Mechanically small:* cracks where the near-tip plasticity is comparable to the crack size or which is encompassed by the plastic strain field of a notch.
- *Physically small:* cracks typically less than about 2 mm in length.
- *Chemically small:* cracks which would normally be amenable to linear elastic fracture mechanics but exhibit anomalies in growth rates below a certain size as a consequence of corrosion fatigue effects.

Indeed, while each of these classifications present some difficulties from an engineering standpoint, the behaviour of physically small flaws can be the most difficult to quantify because it has been demonstrated that these types of cracks can grow appreciably faster than long cracks subjected to the same nominal stress intensity range. The studies of short crack behaviour and non-propagating cracks mean that fatigue failures now have the potential to be explained in terms of crack growth throughout the process from the first strain cycle to final rupture using fracture mechanics principles. Additionally, it seems likely that many familiar classical techniques and empirical factors used to estimate fatigue strength will be superseded in the future; for example, the treatment of multiaxial stress states, notch sensitivity, stress-strain cycle counting and strain hardening or softening characteristics.

A good illustration of this change of thinking is that the fatigue limit was considered previously to be identified only by a stress value. A better concept, however, is the ability of a crack, whatever its length, to propagate to failure because the fatigue limit refers to the stress level required to overcome the strongest barrier to crack growth which will be represented by a microstructural distance.

Apart from the greater susceptibility of soft whitemetal bearing materials to crack, the effects of temperature on the fatigue strength of most engineering materials are not serious until about 400° C when all mechanical properties are progressively impaired. Temperature fluctuations can, however, cause severe thermally-induced cyclic stress gradients which in turn induce fatigue damage which is manifested as crazed cracking. Typical examples are the overheating of diesel engine bearings and consequent dam-

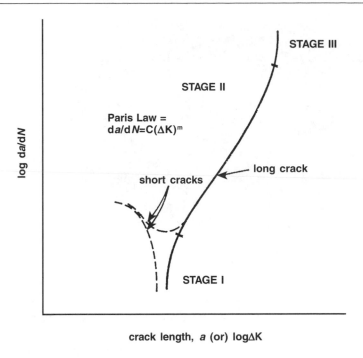

Figure 9.7 *Fatigue crack growth.*

age to crankpins and journals and the impingement of water on the bore surfaces of superheated steam pipes.

Fretting fatigue is a special case of fatigue action which results from a combined mechanical and chemical action and in which three fundamental conditions are necessary for the failure mechanism to develop. These are: the ability of two surfaces to move relative to each other, albeit by a small amount; points of asperity on the surfaces which make contact and the presence of sufficiently high stresses in the vicinity of the contacting points to cause surface cracking. Typical engineering situations giving rise to the conditions promoting fretting action are flat contacting faces such as flanges where shear loads across the faces may exist and the normal forces may permit some degree of slippage between them. Alternatively, holes which, for example, may house bolt shanks or rivets and be subjected to interface sliding and varying interface pressures. Further situations include key and keyway interfaces, leaf springs, splines and contacting strands in wire ropes.

When surfaces are permitted to rub together under the fretting conditions described, scars on the surface tend to form relatively rapidly and these often have a roughened appearance. In cases of fretting in steel the scars contain a reddish-brown oxide, often in the form of a powdery deposit although in some cases this may form a glaze. Alternatively, if the action is between aluminium surfaces then the deposit is black. Research into the fretting fatigue mechanism has suggested that the nucleation of fatigue can

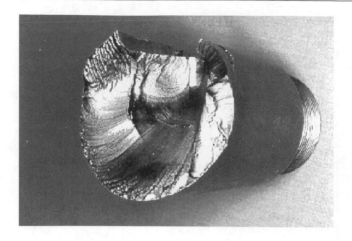

Figure 9.8   *Fretting fatigue of a tapered shaft.*

result from one of a number of causes; typically these are an abrasive pit-digging mechanism, asperity-contact microcrack initiation, friction generated cyclic stresses that lead to the formation of microcracks and subsurface cyclic stresses that lead to surface delamination in the fretted region. It has also been shown that compressive stress between the members can have a beneficial effect on the suppression of fretting fatigue. Much research effort is, however, needed in the subject of fretting fatigue.

Figure 9.8 shows an example of a failure caused by fretting initiated torsional fatigue. This failure was induced by fretting which occurred at the interface between a drive shaft and fluid coupling due to an inadequate compression fit on a 1:10 taper. This fretting action induced over one hundred separate fatigue crack initiation sites on the shaft. In this case the function of the key was to provide a reserve safety factor to the design. The lack of compression fit was induced partly by the high taper and in part by the original design not taking due account of the restricted conditions under which the required compression fit had to be achieved during construction or maintenance in the ship.

Brittle failure

True brittle fracture occurs without significant gross deformation. This fracture mechanism manifests itself in a plane normal to the applied stress and is essentially a fracture mechanism which occurs through the grains of the material. When viewed under the microscope the fracture face is seen to contain a large number of facets together with a branching pattern of cracks. In plate sections a pattern of *chevron* markings is normally seen, the direction of which points to the origin of the failure (Fig. 9.9). While brittle fracture is not a commonly encountered type of failure today, a recent example of its appearance is the failure of some 38 mm bolts (Fig. 9.10) from a steering gear hydraulic ram stopper pedestal. These bolts were manufactured from a 0.35% C, 1.48% Mn steel.

– 142 –

Figure 9.9   *Chevron markings on a brittle fracture of a plate.*

Material tests showed them to have a ratio of the 0.2% proof to ultimate strength of 98% and an elongation of only 9%. The bolts had failed at the underhead position in way of a 0.45 mm radius by brittle fracture. Also around the circumference of the failed section is a small 5 mm deep region of fatigue propagation from multiple origins.

Ductile failures types

Pure ductile failures are rarely encountered today in the marine industry, however, a variant of this mechanism is occasionally seen. This is the lamellar tearing mechanism which is known to sometimes occur in the parent plate beneath welds in the through thickness direction. Figure 9.11 shows a classical example in a cruciform joint. In this

Figure 9.10   *Brittle fracture of a bolt.*

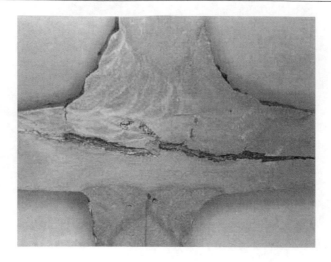

Figure 9.11 *Lamellar tear on a cruciform joint.*

type of tearing mechanism the presence of non-metallic inclusions has a significant affect on its development. Another example has recently been found in a fractured fuel injector. The injector was made from a medium carbon sulphur bearing steel with a significant quantity of manganese sulphide inclusions elongated in the axial direction. The holder had cracked longitudinally due to the action of lamellar tearing caused by high hoop stresses and an unfavourable microstructure of the material (Fig. 9.12).

Some lessons learnt from failure investigations

There are a great many lessons that can be learnt from failure investigation activities such as these. Some are specific to certain areas of engineering application while others are more general in their nature. The underlying general lessons can be summarised as follows:

- When undertaking engineering design it is essential to *stand back* from the detail of the design and look at the whole engineering problem. From such an exercise it is possible, for suitably experienced engineers, to identify the weak points in the design.
- Today the investigator, or designer, has a range of analytical and experimental capabilities at his disposal. In most engineering situations these capabilities only give a partial picture of the problem and, as such, they are aids and not a substitute for sound engineering judgement.
- TID's database shows that in a great many instances problems continue to recur on a periodic basis. Whether this is due to technology moving on in other areas which then reintroduces problems of a similar type in a later age, or the lessons of the past not being effectively passed on to new generations of engineers is unclear

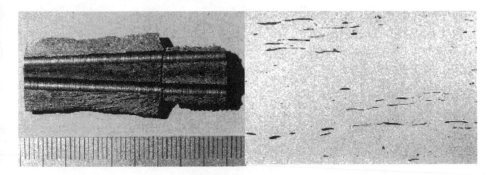

Figure 9.12   *Failure of a injector: Axial fracture face (left); Manganese sulphide inclusions in the material (right).*

    – perhaps a combination of the two causes.
- A great many failures are caused by a lack of attention to the detail design of individual components. Moreover, problems frequently arise when similar care is not exercised in the integration of components into an engineering product or system, especially when different manufacturing sources are involved.
- Similarly, installation, maintenance and operational procedures in relation to the design philosophy are critical if failures are to be avoided.
- Investigators must always attempt to identify and understand the prime cause of the problem so that proper judgement can be exercised in deciding whether fundamental changes are needed or if palliative countermeasures will be sufficient.
- The training of engineers in the detailed practical and theoretical skills of engineering is fundamentally important, but more critical is the combining of these skills into a unified professional engineering knowledge and instinct. Furthermore, within this concept engineers must gain an in depth knowledge and appreciation of the materials with which they work.
- Continuing professional development is an essential ingredient of any engineer's training. It must, however, be implemented rigorously and honestly on the part of the individual concerned and the learned institutions.

    The quality of a design, or an investigation, is directly dependent on the quality and motivation of the people employed in that function. As such, the rewards given to the individuals in terms of creativity, job satisfaction and of course remuneration must be such so as to attract the correct calibre of person. These rewards, however, need to be set within a proper framework of responsibility and accountability.

Conclusions

The investigation of failure in engineering situations relies for its success on the quality of the people involved in the activity together with a combination of skills and technologies. The basic technical skills are those of field measurement and study, metallurgical investigation and theoretical and computational analysis; each providing a part of the solution or extending the knowledge gained from one of the other approaches. Equally important, however, are the engineering and diagnostic skills of the investigator. These latter skills, while relying on a sound knowledge of engineering theory and practice, derive more from the application of logical analysis and reasoning to a problem. In addition, an extensive database of previous failure histories is essential.

References

Report of the Propulsor Committee. 1987. *Proc. 18th ITTC.*
Suresh, S. and Ritchie, R.O. 1984. Propagation of Short Fatigue Cracks. *Int. Metals Rev.* **29**, pp. 445–76.

Acknowledgements
The authors wish to thank the Committee of Lloyd's Register for permission to publish this chapter. Additionally, the experience and knowledge of a group such as TID can only be built up by much careful and hard work from the Department's technical and administrative staff over the years. Consequently, thanks are due to the authors' many colleagues, both past and present, who have contributed to and developed the various investigation techniques and without whom the Department would not exist with its present capabilities.

10 The role of risk assessment in failure investigations

S.B.Tietz

Introduction

All human activity entails some risk. The fact that it is unavoidable is generally recognized, though accepted to varying degrees, depending on the nature of the risk and the perception of those potentially exposed to it. Some will be risk seekers or accepters by temperament while others are risk avoiders. There is even evidence that removal of some risk will cause persons purposely to subject themselves to a new one, suggesting that they seek some kind of undefined risk balance in their lives.

Tolerance of risk varies widely. Thus multiple fatalities will be much less acceptable than when the same number of casualties occur in separate incidents. In theory 'acceptance criteria' are statistically based but there are huge variations, depending on the hazard (Hambly and Hambly, 1994). This is well illustrated by the recent controversy in the UK attached to beef on the bone, where the cost of the remedial measures per fatality was large though the risk was deemed to be minuscule when compared to the risk of death or injury in a motor accident, whether assessed on the basis of accident statistics or based on some notional value attached to human life or injury.

Some risk avoidance aims and even requirements backed by legislation, are thus disproportionate when compared to statistics defining average acceptable losses. In part this disparity may be due to the likelihood that those seeking employment in bodies aiming to control risks in one of various inspectorates or health and safety organizations, will be risk avoiders by temperament and are also the predominant advisers to government on such issues. A vociferous press relying on a larger readership through sensational headlines aids such a process.

Public opinion may thus become distorted, with actual opinion more tolerant than it appears. This resulting divergence in levels of acceptability, real or perceived, makes it difficult to set standards of acceptance and thus to give guidance regarding appropriate levels of accident avoidance. The very word accident implies lack of premeditation though this is ever more rarely accepted in searching for the blameworthy.

The more recent factor when considering risk is not so much the awareness that it exists but the endeavour to quantify it as part of a conscious decision making process. This extends into areas where a lack of acceptable criteria makes assessment vague and criteria of acceptability a matter of personal opinion. The subject is under development and some of its wilder aspirations may well fall out of fashion due to their complexity. Nevertheless simplified procedures are now part of a range of requirements, such as the assessment of risks attached to site safety and health of operatives (HSE, 1992, 1995*a, b* and *c*; CEN, 1996; Rothschild, 1978). Similarly the one time "contingency sum" which featured in many Bills of Quantities and contracts, to allow for unexpected extra costs, are now often replaced by a risk analysis when setting construction budgets.

What is a failure?

'Failure' can be any event causing an unexpected loss. Taken in its narrow sense, the investigation of failures presupposes that they have already occurred. Within the field of engineering there is a further, often erroneous, assumption that the failure is probably physical. As risk assessment predominantly addresses the avoidance of failure, it may at first sight appear misplaced in this narrower context. An investigation prior to construction, which addresses factors which could result in failure, is not however altogether different from one which analyses why a failure occurred, though likely to entail many more variables.

Very occasionally the cause of a failure is immediately self evident and attributable to one or very few identified causes. More frequently several contributing factors are possible and one must evaluate different hazards and on the probability of their contribution in relating cause to affect. The difference in the risk analysis which predates failure, compared to one which follows it, may thus primarily be a matter of scale and numbers of variables.

Taking the broader view, investigation of failures thus includes the assessment of hazards, with the aim of avoiding failures. These are in fact the main aims within health and safety legislation, when this recommends risk assessment.

For present purposes one thus needs to divide assessments into
- those which predate construction and aim at avoidance, and
- those which follow some identified failure.

The nature of failures

In broad terms hazards come in three categories:
- Natural hazards, emanating from the physical environment, such as typhoons or earthquake.
- Technological hazards humanly created, including damage from industrial waste, structural failures, etc.
- Social hazards from within society such as disease (which can also be natural), legislative changes, or contractual matters.

There is inevitably some overlap. Thus cracking in a building is a technological failure but where culpability is an issue, it will be in breach of a contractual obligation and thus also a social hazard.

In engineering, physical failures, in order of their severity, are:
* structural collapse;
* serious structural defects/damage;
* cosmetic failure which impairs serviceability and
* less specific claims alleging some loss.

However, failures involving engineers can and frequently do take many forms other than physical defects, e.g.
* the failure of a development to meet expectations, perhaps of design, useable space or quality;
* programme slippage (e.g. late completion causing a claim, including consequential loss);
* budget over-runs;
* inadequate accident avoidance measures;
* management shortcomings resulting in contractual claims and
* poor contract selection or breaches of contract (which may also encompass any of the above).

Most clauses in a construction contract define requirements aimed to provide safeguards which prevent failure in its broader sense. By the same token a breach of the requirements of any such clause is then the failure. Doran and Pepper (1996) list some of these.

Common features in risk assessment

Risks can be assessed in a variety of ways, varying from a sophisticated statistical analysis of several variables to reading of the sea-weed and backing a hunch (Tietz, 1999). Their form will depend on their aim but they share some characteristics.

All risk assessment requires a list of the hazards, which then need assessment, one at a time, to identify their potential severity, their frequency and those who are (or what is) at risk

Though it may seem self evident that only hazards which could reasonably have been foreseen can be assessed, even that factor has been argued in the Courts, with the fact that failure occurred proposed as sufficient reason for culpability.

Results are likely to become less reliable as the number of variables increases

Hazards which are exceptionally remote, are better ignored even if one lists them, as evidence that they were recognized but deemed to be too unlikely to be evaluated. To include them would distort both the outcome and the scale of the task.

For each occurrence one then has to quantify:
* Its likelihood on some scale, possibly of 0 to 4, ranging from 'very remote' to 'probable', or as a percentage, based on the probability of its occurrence. There is however little point in introducing more apparent precision than the underlying data supports.

- Its severity, normally also in one of a series of categories, e.g. catastrophic, severe, acceptable or negligible.

The risk is then the likelihood × the severity.

The appropriateness of a particular type of assessment depends on the nature of the risks being considered.

Physical failure avoidance

This relies heavily on identifying potential structural weaknesses and considering whether compliance with standard practice provides adequate guidance. This is expanded below.

Contractual obligations

This category covers too many issues for potential breaches to be considered here. Legal advice may often be necessary even to interpret requirements and identify pitfalls. There are however some basic ground rules for avoiding disputes, short cuts leading to error or financial failure. These include the following:
- For contracts to be effective, mutual trust must exist between the parties. Each must aim to abide by the contract.
- The risks placed on any party must be known and quantifiable by them within reasonable limits.
- Those asked to take on a risk must be prepared to accept it and compensation for an abnormal risk must include a premium.
- It is rarely economical (or even possible) for any party to avoid all risk. Failure of the risk bearer would probably outweigh the desired benefit. A risk/cost balance is thus advisable.
- At best, risks should be allocated to those best able to control them, who are adequately compensated for taking them on and are able to bear them.

Health and safety issues

This can be assessed in a variety of ways. CIRIA (Construction Industry Research and Information Association, UK) carried out a study which included a suggested framework for construction risks (CIRIA, 1995). The proposals included the tables here reproduced.

Table 10.1 *Typical scales for likelihood/probability.*

Description	Guidance	Scale	Nominated value
Frequent	Likely to occur frequently, many times during the period of concern (e.g., project duration, life of the building).	4	1:1 year
Probable	Several times in the period of concern.	3	1:10 years
Occasional	Sometime in the period of concern.	2	1:100 years
Remote	Unlikely but possible in the period of concern (e.g., once in ten times the life of the building).	1	1:1000 years
Improbable	So unlikely that it can be assumed that it will not occur or it cannot occur.	0	1:10,000 years

Table 10.2 *Typical scales for consequence.*

Description	Guidance	Scale	Nominated value
Catastrophic	Death, system loss, criminal guilt, bankruptcy.	4	£10m
Critical	Occupation threatening injury or illness, major damage, substantial damages; will exceed contingency; dividend at risk.	3	£1m
Serious	Lost time, injury or illness; damage causing down time of plant; consumes contingency, requires an insurance claim.	2	£100k
Marginal	Injury or illness requiring first aid at work only; minor damage that can await routine maintenance, will only require an apology letter, accommodated as part of contingency or insurance excess.	1	£10k
Negligible	So minor as to be regarded as without consequence.	0	£1k

Table 10.3 *Assessment of risk cost.*

		Risk cost £/year				
Scale points connsequence		4	3	2	1	0
Likelihood	Risk cost probability	£10m	£1m	£100k	£10k	£1k
4	1:1 year	10m	1m	100k	10k	1k
3	1:10 years	1m	100k	10k	1k	100
2	1:100 years	100k	10k	1k	100	10
1	1:1000 years	10k	1k	100	10	1
0	1:10,000 years	1k	100	10	1	0

The relative importance of the risk can then be assessed by multiplying the factors from the relevant scale. Thus a probable occurrence (factor 3) which would have a serious consequence (factor 2) would have a relative importance of 6, out of a possible maximum of $4 \times 4 = 16$. By a similar process the risk costs can be assessed, reducing the cost to an annual basis. Table 10.4 gives a typical example.

Numbers alone are not however a wholly reliable guide to the advisable scale of avoidance measures. Thus catastrophic effects may need abnormal avoidance procedures even if the likelihood of their occurrence is marginal or even negligible but nevertheless possible. Leakage of fissile materials from an atomic power plant is an example of this. A weighting can then be given to the simpler analysis, to allow for defined factors, e.g. the risk of multiple injuries or unacceptable knock-on effects.

Risk avoidance measures must then be assessed for each contributing item, with options for its mitigation or removal considered. If mitigation results, with a lesser risk remaining, the process may have to be repeated.

For risks such as serious injury, assessment based on approximate data such as the above tend to be most appropriate, using a notional cost, based on statistical evidence. Various public bodies, e.g. the Department of Transport and the Health and Safety Executive regularly update the estimated cost per fatality and per injury, defined by grades of severity.

Financial loss

This can arise in many forms and there is much specialist guidance on its minimization. Typically the market place offers facilities for buying foreign currency ahead, to take care of the risk of fluctuations over the period of a contract.

Table 10.4   *Assessment of risk acceptability.*

	Risk acceptability				
Consequence	Catastrophe	Critical	Severe	Marginal	Negligible
Likelihood					
Frequent	Unacceptable	Unacceptable	Unacceptable	Undesirable	Undesirable
Probable	Unacceptable	Unacceptable	Undesirable	Undesirable	Acceptable
Occasional	Unacceptable	Undesirable	Undesirable	Acceptable	Acceptable
Remote	Undesirable	Undesirable	Acceptable	Acceptable	Negligible
Improbable	Undesirable	Acceptable	Acceptable	Negligible	Negligible

KEY: **Description**   **Guidance**
       Unacceptable   Intolerable - must be
       Undesirable   eliminated or transferred
       Acceptable
       Negligible

Estimating construction costs

Risk analysis is a useful aid in assessing the contingency sum to be added to an estimated cost of the works and in comparing options in order to arrive at a 'best buy'.

Some such risks can be quantified directly in money terms. The process is inevitably project specific and not precise. Typically the cost of a potential delay in part of a construction programme can be assessed by examining each contributing factor, the likelihood of the occurrence and the maximum and minimum likely cost if it occurs. A sensible assumption, somewhere between these, is then made. Where sufficient data exist, the analysis can be checked against the statistical likelihood and cost, based on data from wider sources. Where large sums are at risk, more than one evaluation may be necessary to test the sensitivity of a specific item as part of the aggregated risk. Where longer delays cause disproportionately greater consequential costs per unit of time, a range of unit costs can be used.

The aggregate of these estimates provides the appropriate contingency covering risk. One can take this further by assessing the cost of achieving different confidence levels, where 100% represents the contingency required to be reasonably certain that estimated costs plus the contingency will not be exceeded. As near certainty becomes expensive, some more realistic confidence level is commonly taken.

The figures below use the example of a foundation, with the figures used self-evidently site dependant.

Description	Occurrence probability	Minimum estimate	Maximum estimate	Assessed estimate
Hard spots in dig	3%	£1,200	£7,000	£2,300
Polluted areas	2%	£3,000	£25,000	£4,530
Old foundations/obstructions	0%	£0	£4,000	£500
High water table and pumping	15%	£3,000	£8,000	£2,235
Excavations fall in	10%	£1,500	£4,000	£5,320
Soft spots and overdig	20%	£1,000	£4,000	£3,450
L.A/consultant rejects bottom	15%	£2,000	£7,000	£2,860
Overdig made necessary by delayed blinding	10%	£1,500	£3,400	£2,250
Delayed concrete delivery	25%	£2,800	£4,300	£2,830

The figures may include both the direct remedial costs and consequential costs such as delays or these can be assessed separately, the choice being the assessor's. The 'Assessed Estimate' column relies on data held by, among others, many quantity surveyors. It is based on the analysis of similar items, as found on a range of previous contracts and is thus as reliable as the size and analytical quality of the database. Figures from it can fall below the minimum or exceed the maximum estimate, but they provide some cross checks on other assumptions.

Though the aggregate cost of the hazards can be established by simple addition, some adjustments may be needed (using, for example, the least square rule) to allow for the improbability that every hazard will become actual. More sophisticated methods, using standard distribution, are appropriate where precedent provides sufficient statistical evidence. Relevant experience is necessary in such evaluations as the potential costs of a loss can otherwise become unrealistic and sufficiently frightening to inhibit any action.

The risk factor is best declared as a separate contingency within the building budget, partly to avoid heart failure in one's client but also as it is not an entitlement in the contract. It should be possible for risks which arise to be tracked back, to be compared to the allowance made for them. Thus issues of design should be separate from those of construction, and those occasioned by the client (which may be changes in his brief) or the local authority (such as a delayed response to a planning application). This may not change the final development costs but protects parties against wrongful claims.

When considering the impact of each option in a risk assessment one also needs to:
- assess any transfer of risk occasioned by the risk avoidance or mitigation. Typically rerouting of construction traffic to avoid a vulnerable area and possible claims will transfer some risk to another location, perhaps reduce it overall but not remove it. Unless an assessment is specific to one site only, the transfer should not be ignored. Such transfers may introduce other and different criteria. Thus a permanently available cherry picker, which provides access to lights in a tall space

may make access safer but may also take up valuable space. These disparate issues must then be put on a common base, commonly a notional cash value, if options are to be compared. Similarly a saving in construction may result in a component with a shorter life.

• Quantify potential consequential losses as well as those directly applying. For example a developer aiming for a planning consent may wish to improve his chances of obtaining his consent by reducing the footprint of his development. The loss of area and potential rent is a direct loss. If the market is particularly strong for the larger building, there may be the indirect loss of some potential tenants who would lose interest, perhaps resulting in a reduced rental level.

Residual risks

These tend to remain even after all reasonable precautions have been taken. Typically pipe work for drainage is sized to allow for predictable storm flows. This is based on commonly accepted criteria but does not exclude the impact of an exceptional storm. Nor does a sensibly defining soil survey avoid the possibility of a local fault on a site. The adopted option is then a balance between the additional cost and the residual risk. This requires judgment.

Quite complex risk and cost benefit assessments are used when determining best buys for types of public expenditure. Typically this applies to route and capacity evaluation of proposed highways. While such quantification works well when comparing like alternatives, it becomes particularly unreliable when comparing disparate activities, such as issues of sustainability or social issues. Risk assessment may still provide the most appropriate methodology, though largely due to the absence of reliable alternatives but choice of the 'best buy' is readily overtaken by political preferences.

Constraints to reliability

A significant limitation of risk assessments tends to be that any 'best' solution depends on the criteria being addressed. The cheapest may not be the optimum solution. Similarly, an analysis concentrating predominantly on safety may be relatively simple but can be misleading in a wider context, when considering optimum solutions. The safest option which is likely to avoid injury may be disproportionately more expensive and slower to construct; the option giving easiest constructability may provide less than the optimum space or a less attractive building (with the added problem of putting a price on aesthetic considerations) or the fastest programme may require a construction contract where costs are less easily controlled.

Consequential losses often heavily outweigh direct loss in money terms but can be even harder to evaluate, with the quantification of damage often subjective. This particularly applies to pre-construction assessments. Taking an example, if an excavation falls in, an adjoining road may need closure. How does one cost potential disturbance to traffic? Could the water main under the road also rupture? Could a passing bus fall into the excavation and what is then the potential list of dead and injured? Once this same excavation has fallen in, one at least has some facts to go on, e.g. there was no bus and the water main remained intact. Though the courts then look for clear evidence of a loss in their evaluation, this can still be highly subjective, particularly if the claim relates to intangibles such as stress created by noise, a diminished reputation or loss of business. Delays are equally complex, with assessment of cause commonly a mixture of facts and opinions.

If the evaluation of conceivable knock-on effects thus becomes too fanciful, the potential damage could appear to be huge. What could go wrong, including the unexpected opinion of a judge making an award, is then in conflict with probability. That does not suggest that consequential losses can be ignored, but it requires a hard look at the value attached to common identifiable risks, compared to the less tangible ones.

As with all aspects of risk assessment, judgment is therefore important but one has to accept that even good judgment is not necessarily right, any more than an apparently near cast-iron case will necessarily succeed at law.

While most risk avoidance is subjective, this applies above all when considering the acceptability of a risk. In the political arena the impact of any recent mishap such as a rail crash or a health scare will weigh disproportionately heavily on the decision making process for a period. A view, probably influenced heavily by media reaction, will be taken and much more stringent avoidance measures may be demanded than would have seemed justified before the accident, or perhaps later when the accident is no longer to the forefront of public anxiety. The likelihood or improbability of a recurrence weighs insufficiently in such cases.

Self evidently such changes of opinion cannot be forecast ahead of their cause and any attempt to allow for the worst scenario would rapidly lead to bankruptcy. Those who bear the cost are likely to have a different opinion from those who need only to assess, either in order to prevent a failure (e.g. local authorities or HSE) or to determine appropriate post-failure action. The saying that 'no cost is too high for the onlooker who need not pay' is then all too true.

Risk assessment is thus not a general vade mecum, able to cure all ills but simply a useful tool when properly used. Its greatest asset may be that the wider implications of a project are more likely to be considered in good time.

Pre-construction risk avoidance

Risk assessment has not featured historically by that name when making design decisions but its ingredients have always been present.

Risk assessment in design

Many design considerations rely on statistically acceptable norms. Thus wind resistance is analysed on the basis of defined return periods, normally but not necessarily adopting the likelihood of a wind occurring once in 50 years. Similarly rainfall statistics and predictions of run off determine the size of drainage systems. Where the impact of a failure would be unusually severe, as with a large grandstand or a dam, the statistically based return period would be longer. A specific client may also have different requirements and typically design of a building such as a cathedral, with a much longer life expectancy, should then rationally match this by using longer return periods for wind, snow and rain storms. Engineers have long accepted this without necessarily realizing it to be part of a risk analysis.

When assessing the quality or appropriateness of existing structures for a change of use, the extent of sampling undertaken to establish their condition is determined by the evidence we have, based on visual inspection, possibly backed by documentary evidence, and the consistency obtained from the results of initial investigations. The likelihood of localized defects and their identification is then balanced against the cost of further sampling, the damage this creates and the time available prior to making some decisions and proceeding with the works.

For soil investigations a similar process applies, starting with a desk study of geological data, a search for data related to the site or land near it, establishment of previous uses and an examination of the state of nearby construction. That defines expectations which further investigations will hopefully support, and which will then determine the extent of further examinations. In theory the grid used for sampling determines the residual risk but this still does not remove the possibility of meeting unexpected soil conditions. Such norms do however set the basis for our assessments, and the accepted figures in guidance documents.

These are the simpler examples because they are well codified. More often there are further less well codified factors. A soil investigation may define the anticipated settlement but the engineer then still needs to determine what is acceptable, bearing in mind the type of construction, clients' legitimate expectations, bearing in mind the expected use and quality of the construction and the susceptibility of components to movement damage. He must then aim at the optimum solution, with prevention perhaps ranging from larger foundations, with reinforced and unreinforced options compared, deeper foundations if the soil improves at depth or piling in its infinite variety. The acceptability of some cosmetic damage also needs consideration, though best shared with one's client! Are cracks expected to be repetitive or only expected once and thus readily addressed by the first redecoration? How might they influence a sale or lease?

Normally questions of site access, the time required to build each option and costs enter the decision making process. The need for such decisions is commonplace in engineering but there is now an increasing need to demonstrate that adequate consideration was given to options, and that risks attached to each were evaluated. This applies even more where issues of health and safety arise.

The engineer whose client alleges that extra costs or an extended programme arose because inappropriate solutions were selected, could face just as big a claim for damages as when structural failure occurs. This still begs the question how to quantify an unknown risk such as localized bad ground when there may no evidence of it in advance. At best one can identify the work likely to result if such faults occur, cost the consequences and contain them as a contingency or risk analysis.

Design and the contract – Reasonable competence vs. fitness for purpose

The design process requires the selection of the right materials and permissible stresses for each component, with deformation and behaviour in mind. The choice depends on the brief but the engineer would be expected to understand the issues better than his client and thus is likely to be held responsible if he fails to exercises reasonable competence. His requirements may be well enough defined for construction (at least in theory), in a detailed set of contract documents backed by drawings.

Once the purpose is defined, 'Fitness for purpose', which is a normal requirement from the contractor, achieves some meaning. It tends to be less appropriate for design for which 'the purpose' is only implied and thus ambiguous. Did the client ask for storage areas taking heavy loads, or for a long life roof alas only felted? Was it made clear that deep excavations for some item of plant were contemplated close to foundations? Were the accepted norms for assessing natural phenomena such as wind enough or should higher standards have been aimed at? Alternatively was the purpose not achieved because damage resulted in an abnormal storm? Fitness for purpose if accepted, rather than the more normal criterion of reasonable skill and care, thus puts an added risk onto the designer.

Increasingly whole life behaviour will also be considered, influenced by considerations of energy conservation, recycling, environmental impact and sustainability. The evaluation of materials and design factors and the comparison of risks attached to options then becomes even more complex as there is less evidence based on precedent. Progress in such areas requires better statistical evidence of long term behaviour, which will also be influenced by other unpredictable factors, such as the quality of the building maintenance and management. Fitness for purpose then becomes even more of a lottery but that does not prevent its appearance in some design or design and build contracts. What additional design fee then equates to the added risks?

Though meaningful risk analysis considering options is inevitably imprecise, current fashion demands that it should be attempted. It may have genuine benefits in creating a culture, which obliges designers to consider their work in a wider context but will also lead to ever longer documents in an endeavour to define the brief and the assumptions underlying it. In the current climate the identified risks, the criteria used in making decisions and the options considered and discarded should however be documented in case subsequent proof is needed, to confirm expectations and illustrate competence and thoroughness.

Post-failure assessments

Though pre- and post-failure assessments share many features, they self-evidently concentrate on different issues. Post-failure, the identification of cause is likely to be paramount but it may also be necessary to determine whether an adequate assessment in good time could or should have avoided the failure. This will consider many of the issues listed above. Did the engineer accept the constraints of a fixed budget price?

Were the assumptions made regarding natural phenomena reasonable and were they interpreted in accordance with codified procedures? Were other similarly appropriate documents followed in the design? Was the damage foreseeable and was its avoidance possible, using acceptable technical or budgetary norms? These are likely to be a significant factor when determining culpability.

Few failures can be attributed to winds, rainstorms or snow loading which exceeded those to be expected within the accepted return periods. More commonly one thus needs to search further and consider the interpretation adopted for the relevant documents, to establish whether cause was design, construction, manufacture or misuse. Risk assessment (perhaps in that case a smart term for probabilistic analysis) then can assist when there are several likely contributors to failure and their relative relevance is considered.

Proof of an adequately executed pre-construction risk assessment is particularly relevant in claims for injury or ill health, where the recommended procedures are better documented (HSE, 1992, 1995a, 1995b). The intention of various guides backing health and safety legislation is that the potential cause of accidents is at best removed or, if that is not feasible, that safeguards are in place which reduce the risk of accidents. Typically, hazards and their avoidance may include
- falling from a height,
- the practicability of and justification for safety rails,
- avoidance of weak points such as unprotected sky lights on a roof,
- bollards alongside access roads or
- the reduction of unit sizes in construction, to reduce the risk of strains for those handling components.

Considerations include maintenance and demolition procedures. An engineer may then need to prove that he addressed such issues and that his choice was well reasoned.

Alternatively the aim may be to quantify the best course of action for a damage limitation exercise. What is the best case which can be supported by the evidence and is that deemed to be sufficient for a defence? What is the worst scenario? Alternatively, even if a defence can be mounted, what may be the downside? Typically in a civil action that may take the form of excessive costs of pursuing an action, delays to a settlement, 'not worth powder and shot', excessive commitment of time by key personnel, estrangement from a hitherto good client or adverse publicity which outweighs other potential gains. Where one party can draw on disparately greater resources, this too should give pause for thought.

Risk assessment in litigation

All claims start with the desire for compensation for what the claimant believes is a loss which he can substantiate. Some less realistic claims may still be pursued, knowing that the defendant, even if poor and/or perhaps only marginally to blame, will have an insurer standing behind him. In part the lack of 'no fault' insurance is to blame for some claims which might have been more readily settled by negotiation but our culture has also moved towards a 'search for the guilty', following American precedent. Inevitably a desire to avoid risks or at least culpability for them increases proportionately and may be responsible for the impetus to deploy more rigorous risk analysis. Without doubt it now receives widespread attention from diverse sources as the reading list shows.

In the courts the argument will often hinge on what the reasonable person applying reasonable skill and care should know and anticipate. In the case of an engineer this is likely to be judged by the standards to be expected from his peer group. The criteria are not rigid, hence the frequent differences of opinion among experts. An assessor judging between their opinions may take a yet different and perhaps 'unusual' view.

Additional considerations become relevant when contemplating litigation. A claim for compensation now starts with the claimant's 'letter of claim'. It only leads to full litigation when the procedures resulting from the Woolf reforms have been met. These include a requirement for disclosure, usually meetings of both parties and their advisers and the opportunity for the defendant to respond to or reject the letter of claim, with reasons given. The procedures also need to comply with a rigid timetable.

The claim needs to be quantified and alleged direct losses can be significantly increased as a result of alleged consequential losses. Both plaintiff and defendants need to be realistic in assessing their position. Many claims start with some delusions by plaintiffs regarding both the extent of damage suffered and potential chances of recovery. They should aim to avoid excessive optimism. All potential litigants should regularly remind themselves that between 50 and 80% of all court awards are eaten up by costs. Uncertainty is also endemic to the system.

If there is a fatality, serious injury or some endangering to health, the claim is usually pursued in the first case by the Health and Safety Executive. This can create more onerous (though not necessarily more expensive) problems than a civil claim as the prosecution may well be under criminal law which is not insurable or which at best only has the litigating costs covered by an insurer. Also such claims are often followed later by civil claims on behalf of dependants.

Unfortunately, when considering prosecution, the fact that an accident has occurred tends to be taken as a proof of inadequacies. Reasons for not taking some identifiable preventative measures thus need to be well documented. There may be a need to prove this on the basis of 'assessed risk as low as reasonably practical' – ALARP, as against the perhaps less stringent and often more realistic basis of 'best available technology not entailing excessive cost', i.e. BATNEEC. Alternatively the risk may have been deemed to be too remote, though this may be argued against, except in extreme circumstances.

Too many civil claims are pursued and then lost because expert opinions address favourable side issues and lose sight of the main issues, which may be less favourable. Typically one case was taken to court because some concrete, which was faulty on all the evidence, was deemed by the defendants' technical advisers to be wrongly assessed because of minor imperfections in the cubes and the conditions under which they were tested. A realistic appraisal should have recognized that these factors would at most have underestimated the strength by about 5% whereas the strength of the concrete was more than 20% below requirements. A claim which could in the early stages have been settled for perhaps £100,000, thus became a loss, with the final costs of defending it well in excess of £1 million.

Some claims are pursued because some party, often within the claimant's organisation, is reluctant to disclose his or his colleagues' shortcomings, particularly if public investigations, loss of employment or curtailed future promotion are feared. When possible all parties should thus run the full extent of their case, warts and all, past their board or the most senior members of their administration, preferably choosing those who had no direct involvement in the alleged causes of the claim. A preliminary appraisal of the legal and technical position may also be advisable from an independent source and is relatively cheap when compared to the costs of then pursuing an action.

All parties also need to appraise the potential litigation costs and the possible further loss through delays – most cases used to take three or more years to reach a hearing though strict timetables, now set by the courts aim to shorten this significantly – adverse publicity and loss of reputation. Mitigation of a potential loss is then relevant.

Considerations to be addressed as part of an early appraisal should include:
- what form of dispute resolution is most appropriate,
- what is the likelihood of success at law and
- how might potential losses be limited?

Self-evidently potential claimants will only pursue matters if they believe that they have an adequate case or, taking the cynical view, if they feel that they have nothing to lose. The Woolf Report aimed to reduce costs, delays and complexity of litigation and show fairness to all parties. Among the changes made, litigants are now likely to have early knowledge of the case and its defence, will have access to all parties' written evidence earlier in the dispute and the courts will be more pro-active in case management. It is thereby expected that claims may in future be more realistic and better prepared from the start.

Procedural options

The type of hearing is important. While small claims are dealt with different and well-defined procedures, most larger construction claims brought under civil law are heard in the Technology and Construction Court, previously known as the Official Referees Court. Civil claims arising from fatalities or injuries and matters related to owners and tenants' disputes may be heard in the Queen's Bench Division. Alternatively the parties may agree that the case should go to arbitration, with the number of parties then involved commonly but not inevitably limited to one plaintiff and one defendant. No juries are involved in civil claims.

Whereas arbitration used to be considered as the cheaper option, with the possible additional advantage that the arbitrator would be an expert on the themes in dispute, the procedures have increasingly come closer to those used in the High Court, with similar numbers of legal and technical advisers involved and costs not significantly less. The different procedures, which favour one type of hearing against another, require expert legal advice and are not considered here.

A large number of further options, collectively known as alternative dispute resolution procedures (ADRs) are available. These aim to achieve an out of court settlement and include conciliation, mediation and adjudication, each of which normally relies on a single person with adequate experience and background knowledge to assist in a settlement. The difference between them is largely one of the power given to that person to affect a result. He may listen to the parties, separately or together with the aim that with persuasion they reach a settlement (conciliation or mediation) or he may recommend an award (adjudication), with the terms of his appointment determining the extent to which this is to be binding. The details are not relevant here but can be found in a useful book on the subject (Campbell, 1996).

The common factor to ADRs is that the assessor cannot normally impose a settlement and they thus rely on a genuine will by the parties to reach a settlement. They require less detailed preparation, reduced paper work, usually lead to a much shorter hearing and therefore lower costs. The option of accepting the decision of such a hearing or having the ability of taking it on to a further court hearing is usually agreed in advance by the parties but each requires a will to compromise, without which such procedures are unlikely to produce results and will only be a further stepping stone to a court hearing which will then impose such a settlement.

Once a letter of claim is served and defined, the case for defending it can be determined. The potential costs of defending it in court as against seeking a settlement then have to be evaluated. In the early stages the primary costs will arise from solicitors' advice and experts' opinions on technical issues. Such costs, though not inconsiderable, are very much less than those of the same parties in defining and progressing a detailed case. In due course counsels' costs will come into play. If both senior and junior counsel are engaged, this raises costs, particularly as the start of a court hearing nears.

Procedural rules set by the court require disclosure of all documents, which are listed, with the lists exchanged for parties to identify those required. This can be expensive and now occurs earlier in the pre-trial timetable. Often more than two parties become involved, either as defendants or as third parties, joined in the action by the first (or sequential) defendant, with each party likely to run up proportionate costs.

There is thus a significant financial incentive to achieve a settlement prior to reaching the courts, particularly in the early stages. Too many cases fail to settle because the costs have become too heavy for the parties to meet them. Both hope for a win and the losers may well be unable to meet their obligations. If an action goes all the way, only the winner is likely to have his costs met by the other parties and even he is unlikely to be reimbursed in full. As the costs in running an action can very readily exceed the initial claim, perhaps by a wide margin, the downside of losing is dramatic.

Legally-aided cases provide special problems. A defendant who wins his case, even if the victory is total, is unlikely to have most of his costs met. The victory may thus be pyrrhic in terms of its financial outcome though it may save a reputation.

It is difficult to quantify the cost of bringing an action and thus identify comparative risks. The claim as such is quantified when the letter of claim is served, though then frequently modified in the light of further evidence and discussions. Legal and technical opinions are commonly required. They may require advice in various fields typically, respectively covering issues of cause, costs and programmes. More than one of a discipline may be necessary if specialist issues arise.

As a very rough guide, and at rates relevant in 2000, lawyers' charges in central

London are likely to exceed £1,200 per day (commonly charged in 6 minute units and then aggregated) and each expert's costs may well exceed £700 per day. For expert advisers or experts the predominant cost arises from the examination of all the evidence and thus the arrival at an opinion, adequately backed by reference to the facts on which it is based. Meetings between experts for the different parties may also prove time consuming though probably cost effective if they achieve their objective of identifying and limiting technical differences and thus shortening the trial. Most of the experts' costs tend thus to arise prior to a court hearing, some of them soon after a letter of claim is served. Barristers' costs depend on their seniority and are commonly split into an initial charge for an opinion, fresheners at defined stages and then per diem costs in court.

The actual hearing may cost an average of around £1000 per day per professional in court and for a full court hearing of a substantial dispute, with junior counsel and a QC involved, there are likely to be at least five persons representing each party on an average day. Witnesses of fact may increase that number though both they and the experts will only be present at the hearing intermittently. Pre-trial reviews before a judge will give an indication of the anticipated length of a case, with the number of intangibles often proportional to this.

For most cases a competent solicitor will be able to give early estimates of anticipated costs long before the trial. Rough figures of the kind quoted above may however assist in the very early stages, when potential participants in litigation need to assess the risks of proceeding to an action or effecting a settlement. Such settlements are also more easily reached before entrenched positions are disclosed and heavy costs have been incurred.

Once a case is under way there are still options for damage limitation, for example if both parties agree a settlement, each paying his own costs. Judges may well try to assist some conciliation at pre-trial reviews. Offers of payment into court (with details known to the parties but not declared to the judge) can place a limit on damages awarded but pre-offer costs are not covered by such payments which thus normally only apply when considerable costs have already been incurred.

Going through the stages of preparing/evaluating a case and the hazards can be examined at each stage.

Stage 1

Before a letter of claim can sensibly be issued, a potential claimant needs advice on the strength of a potential case, both technical and in law, and the procedural options. Can a claim results be adequately defined, is it realistic, what is its nature and what form of hearing is likely? Are proceedings civil or may criminal proceedings arise?

Stage 2

Following the letter of claim the defendant needs to consider his options. Is the alleged failure accepted in full, only partially accepted or denied and are the causes adequately defined? Do the facts support a plausible claim/defence? Have insurers been informed and what is their attitude?

Documents need to be disclosed by all implicated parties. Advisers meet and exchange opinions on the total evidence, determining merits and demerits of the case. Unless unbiased opinion is available in house, independent advice is required. The advisers will determine whether total or partial agreement exists on aspects of the case, i.e. they aim to identify and narrow the issues and, where disagreement remains, define the differing opinions. There should thus be a clearer picture on each party's claims, their strengths and weaknesses. Before formal litigation is contemplated, both plaintiff and defendant need to reassess preliminary legal and technical opinions on the strength of their case, the negative aspects and the potential for claims avoidance.

- Are there important side issues (e.g. adverse publicity, tying up of essential staff to pursue the case or loss of clients)?
- Can these risks be reduced/influenced and what are the implications of doing so?
- Is time required to raise necessary resources if a loss is likely?
- Are further third parties likely to be joined in the action and if so, by whom?
- What would be the impact on in-house personnel of pursuing the claim?
- Have they the knowledge and time to provide the back-up which is an essential contribution to solicitors and experts in the preparation of a case?
- Would the profits they might generate in pursuit of their normal work exceed the possible recovery in a claim?
- Would they be credible witnesses of fact if called on to provide evidence?

Stage 3

If the dispute is to be pursued, the court needs to be advised and told that experts are required. The court must consent to this before they may be called on to provide an opinion to any proceedings. They may be but are not necessarily the advisers who considered the claim. Whereas expert advisers have a responsibility to their client, experts (whether previously giving clients their advice or newly appointed) have a duty only to the court. The solicitor's letter of instructions, which is disclosable to the court will then brief them and they require full access to all evidence to establish whether they can still support the proposed case. Opinions on the causes of the failure, those responsible and some indication of the extent of their responsibility should by then be clearer. If expert opinion is at that stage unfavourable, it may be the last opportunity to change the pleadings, change the experts or seek an early settlement.

Further information required in the form of site investigations tests or other in depth studies needs to be identified, budgeted for, and agreed with other parties if findings are to be shared. Such requirements also need to be controlled, to ensure that the scale of the dispute justifies their cost. Once committed, experts are obliged to assist the court and must give an unbiased opinion, warts and all. It could prove very dangerous to withhold information from the expert or by him and then be surprised by its exposure in court.

The timetable and its implications pre-trial and at trial needs to be sketched in, to be confirmed at pre-trial reviews, presided over by the judge. The Woolf reforms make such timetables more demanding.

Lawyers and the client need to determine whether to seek an ADR at any stage and if so, what type.

Stage 4

Once writs are served, pleadings and/or the defence need to be finalised, including counterclaims, together with requests for further particulars where the claim or the defence need amplification to be fully addressed. Even for an ADR, details of the case and the pleadings/defence need to be prepared, though in less depth, but backed by a core bundle of supporting papers, made available to the other party and the person conducting the hearing.

Stage 5

Experts need to meet again, in all probability as decreed by the court, with the aim of achieving a joint report which sets out points of agreement and differences. The joint opinions of experts are privileged (i.e. they may not be used in court) until all parties agree to their release, when they become 'open' documents. They must prepare their written opinions for final review, prior to serving them for exchange. Once the opinions of the various experts are known, with reasons given where they cannot agree, a clearer view of the likely technical merits of a case should be evident. This may be the last opportunity to effect a settlement before more heavy expenses arise.

Stage 6

From there on the lawyers will conduct the case up to and during the hearing but the client has only limited control on mounting costs unless he aborts. He will still remain heavily involved in monitoring progress and ensuring that presentation of the case is truthful and as requested by him but risk assessment is by now of limited relevance. The die is largely cast.

References

Campbell, P. (ed.) 1996. *Construction Disputes –Avoidance and Resolution*. Caithness, Whittles Publishing.

CEN. 1996. *Eurocode 1 - Basis of design and actions on structures*. 9th Meeting of CEN/TC250/ SC1 25.26 March 1996. Agenda item 9c:ENV 1991-2-7: Accidental actions. Draft January 1996.

CIRIA. 1995. *Control of Risk: a guide to the systematic management of risk from construction*. Funders Report FR/CP/32.

Doran, D. and Pepper, M. 1996. Expert Witness, in *Construction Disputes – Avoidance and Resolution*, (ed. P.Campbell). Caithness, Whittles Publishing.

Hambly, E.C. and Hambly, E.A. 1994. Risk Evaluation and Realism. *Proceedings of the Institution of Civil Engineers*.

HSE. 1992. *Management of Health and Safety at Work Regulations 1992: Approved Code of Practice. L21*. Health and Safety Commission. HMSO.

HSE. 1995a. *A Guide to Managing Health and Safety in Construction*. HMSO.

HSE. 1995b. *Designing for Health and Safety in Construction : A Guide for Designers on the Construction (Design and Management) Regulations 1994*. Construction Industry Advisory Committee. HMSO.

HSE. 1995c. *Managing Construction for Health and Safety: Construction (Design and Management) Regulations 1994. Approved Code of Practice L54*. Health and Safety Commission. HMSO.

Rothschild, Lord. 1978. *Risk*. The Richard Dimbleby Lecture, BBC.

The Engineering Council. 1993. *Guidelines on Risk Issues*.

Tietz, S.B. 1998. Risk analysis – uses and abuses. *Institution of Structural Engineers*, **76**, 20; Discussion, **77**,12 (1999).

Further reading

McDowell, B.D. and Lemer, A.C. (eds.) 1991. *Uses of Risk Analysis to Achieve Balanced Safety in Building Design and Operations*. Committee on Risk Appraisal in the Development of Facilities Design Criteria, National Research Council. Washington DC, National Academy Press.

Menzies, J.B. 1995. Hazards, Risks and Structural Safety. *The Structural Engineer*, **73**, 21.

Neale, B.S. 1995. Hazard and Risk Assessment for Constructions: a regulator's view. *The Structural Engineer*, **73**, 22 .

Roberts, L. 1995. The Public Perception of Risk. *Royal Society of Arts Journal*, **CXLIII**, No 5465.

Thompson, P.A. and Perry, J.G. (eds.) 1992. *Engineering Construction Risks: a guide to project risk analysis and assessment implications for project clients and project managers*. An SERC project report. Thomas Telford Services Ltd.

Ward, S.C., Chapman, C.B. and Curtis, B. 1991. Risk Management. On the Allocation of Risk in Construction Projects. *International Journal of Project Management*. **9**, 3.

Warner, Sir F. (chairman) 1992. *Risk: Analysis Perception and Management*. Report of a Royal Society Study Group. Royal Society.

Zuckerman, Lord. 1980. The Risks of a No-Risk Society. *Year Book of World Affairs*. The London Institute of World Affairs., Stevens and Sons.

Appendix A – Some recurring risks

Inadequate or misunderstood brief

- Unproven design
- Inadequate site data
- Faulty analysis
- Poor budgeting
- Poor programming
- Communication breakdown
- Faults in new concepts
- Materials faults
- Supply problems
- Inappropriate skills
  - in office
  - on site

Inexperienced designer/contractor/sub-contractor/supplier

- Delayed approvals
- Bankruptcies
- Deficient management
  - pre-construction
  - during construction
- Poor quality control
- Inappropriate construction contract
- Misallocation of responsibilities

11 Failures and vulnerabilities of reservoir control structures

Jack Lewin

Introduction

This chapter deals with reservoir control structures. They are safety critical, that is, in the event of failure they can cause a breach of a reservoir or an inadvertent release of substantial masses of water which can result in a surge wave in a river and sudden inundation. Loss of life due to such events is a possibility.

Spillways at reservoirs release flood flow into a river and are used to reduce the level of the reservoir in an emergency, such as seepage, flow through the impervious core of the reservoir or a breach caused by an earthquake. Fixed weirs can serve this purpose. Because they have no moving parts they are the most reliable means of providing discharge but they require long crests and when the flood flow is high, long weirs often cannot be accommodated or require extensive protection against erosion of the dam face if it is subject to overflow. In an earth or rock face dam this can require additional work. Spillway gates (*see* Figs. 11.1 and 11.2) solve this problem because the weir crest can be set much lower than the retention level of the reservoir and the flood flow can be concentrated over a relatively narrow width.

The most advantageous type of gate is radial. It consists of a curved skin plate with the convex side facing the reservoir. The skin plate is reinforced by vertical or horizontal beams and is supported by two sets of gate arms pivotally mounted on trunnions at the piers. The radius of the skin plate is struck at the trunnions. This ensures that all forces due to the thrust of water pass through the trunnions and result in no forces tending to raise or lower the gate.

The design concept is for the spillway and the gates to be capable of passing the probable maximum flood, or at least half of this discharge. When designing the gate installation, many engineers assume that one of the gates is out of operation due to maintenance or an operational defect.

A spillway can be used to draw down the level of its reservoir to the level of the weir crest, which is significantly lower when a gated spillway is used. In an emergency, such as water seepage through the dam or at the abutments which can result in a piping failure, it may be essential that the reservoir level is lowered further than the weir level. A piping failure is the gradual, sometimes rapid, erosion and enlargement of a seepage

Figure 11.1 *Spillway radial gates of the Kotmale Dam in Sri Lanka.*

Figure 11.2 *Spillway gate of the Kotmale Dam in Sri Lanka.*

path which can lead to further damage, or even failure of a dam. At best it will result in loss of water stored in the reservoir.

The bottom outlet permits drawdown of a reservoir. It consists of an inlet at low level in the reservoir, protected by a gate of the vertical lift type (Fig. 11.3), a tunnel frequently routed through the dam abutment and a controlled outlet. The control of discharge may be a valve or a gate. The gate can be of the vertical lift type, of which there are a number of special types, or less frequently a radial gate. At many reservoirs the bottom outlet is formed by the tunnel (Fig. 11.4) which was constructed to divert the river feeding the reservoir while the dam was built. For maintenance and emergency purposes bottom outlet gates are backed by a standby gate, and discharge valves by a butterfly valve.

Reservoir control structures can fail due to structural collapse, which would cause the release of large masses of water. Inadvertent opening or too rapid opening, controlled or uncontrolled, can cause a surge wave to travel down river. A surge wave which remains within the banks of a river can endanger river craft, bathers or fishermen, and in tropical countries, livestock, washerwomen and bathers. If the surge wave spills over the banks, the hazard to life and property becomes much greater.

Figure 11.3 *Vertical lift slide gate for the control of the bottom outlet of a dam.*

Figure 11.4 *Tunnel gate for a bottom outlet of a dam at the fabricator's works (courtesy Ishikawajima Harima).*

The other danger is failure to operate. At spillway gates it can result in overtopping of a reservoir causing erosion of the crest of a dam, or the downstream face of an earth or rockfill dam. The integrity of the dam could be at risk.

Failure of the intake gate to close in the event of a tunnel break can be equally catastrophic. The final discharge gate or valve at the bottom outlet is the safety device if the reservoir has to be drawn down below the spillway weir level in an emergency. It is one of the critical moving structures associated with reservoirs.

Any event or occurrence which may prevent the closure or opening of any of the gates of a reservoir, be it for water supply, irrigation or hydropower generation, can be a risk to life, property and the environment.

In the last decades, dam safety has been re-examined and extensive technical literature exists on the subject. Statistics of dam failures have been collected and analysed (ICOLD, 1995). Corresponding investigations into the hazard and reliability of reservoir appurtenances are more recent. There is a greater awareness that the integrity of a dam installation includes the reliability of gates controlling flood release and the facility to empty a reservoir if a fault develops.

In an analysis of causes of embankment incidents and failures, according to US-COLD (USNRC, 1983), 2% of 240 dams experienced malfunction of gates. Since the publication of this analysis a few catastrophic events have been recorded involving spillway gates and a number which resulted in a risk.

Events at spillway gate installations

In 1967 a spillway gate on the Washi Dam in Japan collapsed suddenly (Yano, 1968). The gate was 12 m high and 9 m wide. It was swept downstream. The cause was dynamic instability induced by eccentricity of the trunnion bearings (Ishii *et al.*, 1977 and 1979).

Eccentricity of the trunnion bearings is sometimes introduced at large radial spillway gates in order to reduce the hoisting force. The resultant of the thrust on the radial face then passes through origin of the radius and if this is above the trunnions there is a couple tending to raise the gate, reducing the hoisting force. The event at the Washi Dam demonstrated that it can also cause a dynamic instability. The practice of eccentric trunnion has been used since 1967. Were the engineers who designed gates after that date aware of the possible vulnerability of the arrangement to a catastrophic failure? Did they study the technical papers which described and analysed the collapse of the Japanese gate?

On 17th July 1995 spillway gate 3 of the Folsom Dam on the American River in Sacramento County, California, collapsed and released a flow of approximately 1130 m³/s to the Lower American River (Bureau of Reclamation, 1996). The gate was

Figure 11.5 *Collapse of a spillway gate of the Folsom Dam, California, releasing an uncontrolled flow of 1130 m³/s (after Bureau of Reclamation, 1996).*

15.24 m high and 12.8 m wide. The failure occurred when the reservoir was nearly full (Fig. 11.5)

At radial gates the angular movement of the trunnion bearings is limited. Articulation of the gate of 45–55° will normally cover the gate movement from the shut to the fully open position. The loaded side of the bearing will be opposite to the water side and for easy access the grease nipple is located on top of the bearing, an unloaded part of the bearing. In this arrangement, the flow of grease will be mainly towards the unloaded side. At the Folsom Dam gates, the loaded side of the higher tensile strength steel trunnion pins had corroded due, in part, to lack of lubrication. The trunnion friction had significantly increased over time.

Collapse occurred when a strut brace in one of the radial arms sheared at its connection (Fig. 11.6). Fortunately there were no casualties. All over the world there are possibly hundreds of spillway gates of this type with similar design of the trunnion bearings. Dam operators have since investigated the bearings of their control gates to check whether a failure could result and replacements have been carried out.

The possibility of failure had existed for a long time and could have been predicted; however this was the first case. Were there engineers who recognised that the design

Figure 11.6 *Collapsed spillway gate of the Folsom Dam, California (after Bureau of Reclamation, 1996).*

and choice of materials of the bearings were vulnerable but were dissuaded from altering it by the numerous installations which had operated apparently satisfactorily?

Following the collapse of spillway gate No. 3 at the Folsom Dam, dam engineers have examined trunnion bearings at older radial gates to ensure that a similar breakdown does not occur. However, recently the design was repeated at the four spillway gates at a high dam in Turkey. When the design was questioned and the event at the Folsom Dam was mentioned, the response was '...we have a number of similar installations at our dams and they have operated satisfactorily for many years.' Sometimes it requires a second disaster before a risk is consistently recognised.

A spillway gate on a Swedish dam collapsed due to debris accumulation. Also in Sweden, a serious breakdown occurred during the remote control of a sector gate (Lagerholm, 1996) due to the gate passing the upper limit switch. The bolts on the gate bearings sheared, causing the gate to break loose and to move down the spillway.

Limit switches can be a vulnerable element in any hoisting installations. The failure of the Swedish gate is not an isolated example, although it is one of the more dramatic ones. The basic precaution when any one engineering function is vulnerable and can result in serious consequences is to introduce redundancy, simply, to double the limit switches and to design the electrical circuit so that a warning signal is displayed when the primary limit switch fails to trip. Redundancy is frequently provided by engineers – but why not at this critical installation?

A spillway gate malfunctioned in 1992 at the Tarbela Dam Pakistan (Khan and Siddique, 1994) when it became stuck during a lowering operation. It fell down, breaking two hoist ropes, damaging the gate and the weir. The gate was 28.6 m high and 15.2 m wide. Over a long period, the clearance between the side sealing plates on the piers (the seal contact plates) and the clamping bar securing the rubber seal on the gate had deteriorated. The cause of the dimensional change was not reliably established.

One instance of failure of a dam due to inability to open the spillway gates occurred in Spain.

The collapse of the gate of the Washi Dam is the only recorded case of complete failure of a gate due to vibration, but there have been many potentially serious cases of gate vibration. Frequently, design features which can cause gate vibration are reproduced when assumptions are made how they will perform in service. Hydraulic gates control the flow of water; hydrodynamics are, therefore, the starting point of design, not the structural or mechanical and electrical features. In many engineering structures the secondary effects can be ignored until an increase in size or loading or service conditions is changed. At water control gates exciting forces due to water flow are present, such as eddy shedding from the lip of the gate, fluctuating flow because of gaps in the sill or the side seals, unsteady flow through narrow openings or intermittent flow re-attachment at structural members or badly designed gate lips. Damping forces are also present, due to seal and trunnion bearing friction and a small amount of structural damping. If the exciting forces become larger due to increase in size or loading or velocity of flow, these forces can become dominant.

The frequently heard adage, 'we have done it before and it worked', can be a dangerous fallacy.

There can be a number of different causes, the design of the gate lip, eddy shedding from the lip of a gate, seal leakage, unfavourable approach flow. At a large multigate sluice installation, self-exciting wave oscillations occurred in the upstream basin when six openings discharged while four others were closed by gates (Novak, 1984).

Incidents and failures of bottom outlets

There are a number of research papers concerning hydrodynamic problems which have occurred at bottom outlets. Because high velocity flow is experienced at bottom outlets compared with spillway gates, hydrodynamic problems are more frequent.

Bottom outlets have failed to open due to silting. At the Barasona reservoir in Spain (Romeo, 1996), the silt had extended to a depth of 20 m adjacent to the dam and had completely blocked the outlet. The problem became dangerous following a major storm in 1993.

A number of bottom outlets are never, or rarely, exercised. A recent survey of reservoir appurtenances at dams in Indonesia identified a number of bottom outlets which had not been operated since impounding of the reservoirs. These are not isolated cases. Lagerholm (1996) noted similar failures to exercise bottom outlets in Sweden. Seals under high pressure are subject to contact welding after some time. Gates and bottom outlets which have not been regularly moved may be difficult or impossible to raise.

A comprehensive survey of the operation of bottom outlets at 50 large dams was carried out in Romania (Ionescu *et al*, 1994). While it may not be representative of experience in other countries, significant deterioration, incidents and failures were recorded. Damage had occurred at 38 gate installations; 60% of the incidents and failures were due to vibration problems, including two structural failures, which occurred after 8 and 20 years operation. Four events of intake clogging made the bottom outlets unavailable and nine vibration problems were classified as 'serious'.

Because bottom outlets are subject to high velocity flow, cavitation damage has occurred. The most destructive happened at the Tarbela Dam on the river Indus in Pakistan.

Cavitation is a man-made phenomenon. It does not exist in nature. It can occur in high velocity flow. Under any circumstances in which liquid pressure is reduced, the formation of gas filled or vapour filled bubbles is possible. When a vapour filled bubble forms it will start to grow but as it is carried along in the moving fluid it will reach a zone in which the pressure is above the vapour pressure. The vapour in the bubble then very rapidly diffuses back into the liquid and the bubble collapses. The resulting jet impinges on the boundary with intense force. The effect can be highly erosive and can cause severe destruction.

In 1974 No. 2 diversion tunnel of the Tarbela Dam collapsed due to cavitation damage (Fig. 11.7). Tarbela is the highest and uppermost dam on the Indus, which is fed by the snowmelt of the Himalayas. During the latter stage of dam construction, tunnels 1 and 2 carried the river flow through their temporary intake past the dam. Tunnel 2 subsequently collapsed. Each of the tunnels was controlled by three vertical

Figure 11.7 *Collapse of No. 2 diversion tunnel of the Tarbela Dam, Pakistan, caused by cavitation due to one of the tunnel gates becoming stuck on lowering.*

lift gates moving on rollers recessed in gate slots. Each gate opening was 4.1 m wide by 13.7 m high. During reservoir filling, a sequence of simple progressive closure was to be carried out. However the filling procedure became more complex due to an intention to test the efficiency of the seepage arrangement and the hydraulic regime changed as a consequence.

The two side gates of tunnel 2 were first closed and during the attempt to close the centre gate it became stuck in the 8.5 m open position (13.7 m equalled fully open). While repeated unsuccessful attempts were made to close the gate over a period of 17 days, the reservoir head steadily increased up to a maximum of 107 m.

The gates in tunnel number 1 were opened to lower the reservoir level but each of the three gates could be opened only partly during the emergency drawdown of the reservoir.

Cavitation occurred due to sheared high velocity flow which is generated when a high velocity stream of water impinges on slowly moving water. The cavitation continued unchecked for 17 days, first demolishing the concrete piers separating the three fluidways and then disintegrating substantial parts of the concrete tunnel linings, resulting in the collapse of the tunnel. It was one of the most destructive failures due to cavitation.

When the reservoir was empty it was found that the rails on which the wheels of the gate moved were missing in the lower section of the gate slot and were the cause of the gate sticking.

The limited opening of the gates controlling the flow into tunnel 1 was due to structural failures in the gate track systems.

The cavitation damage was investigated at Imperial College and presented in a paper to the Institution of Civil Engineers (Kenn and Garrod, 1981). During the discussion a number of comments were made by the speakers which have a wider implication than the Tarbela tunnel collapse. Some of these were:

- The most important lesson to be learned from the tunnel failure is that the design engineer must ensure that owners and operators are aware of the potential dangers of incorrect operation of their design.
- It was never intended that any gate would be held in a part-open position under any head.
- Perhaps the general lesson of Tarbela is that the damage that can be done by water is much greater than seems possible, until it actually happens.
- When operational changes were made, were the views and wishes of the designers given less credence than they deserved? If so, how did this come about?
- The design of Tarbela is inspired by known solutions which have been successfully used on other power plants for other conditions of pressure and water velocity. When is extrapolation of a known design to other hydraulic conditions acceptable?
- Any new design should be re-thought for all the possible situations which may occur during construction and/or under running conditions.
- History shows that if it can be operated, it will be operated; the engineering profession must seek ways of improving communications between designers and operators.

Control system failures

Incidents of inadvertent operation of gates under automatic control have occurred. Rajar and Rryzanowski (1994) record the self-induced opening of spillway gates on the Mavcice Dam in Slovenia. Two radial gates 20 m high and 13.5 m wide opened, discharging at a rate of 1192 m^3/s, equivalent to a 50-year return period flood.

Other incidents of uncontrolled gate openings have occurred but have not been recorded because they have not resulted in loss of life or damage.

A number of gate designers and reservoir operators require that automatic gate control systems are backed by hard wired electrical circuits which inhibit the time of operation of gates or the distance travelled following a command to move a gate.

Common cause failures

Common cause failures, that is failures which affect the operation of a system, are the most serious risk. They range from failure of the mains supply and the back-up system to fire, explosion, earthquake and failure of central control systems.

In electrical installations associated with spillway gates, redundancy is usually provided for transformers, mains switches and supply cables. Standby generating plant is almost invariably provided, either of the permanent or mobile type. Portable plant forms a second standby at some installations. Surprisingly, double busbars are rare in the distribution of electrical power. The failure of a single busbar can be a common cause fault, or at least affect several gates in a multi-gate installation.

The same degree of redundancy in the electrical mains and distribution equipment is rarely provided at bottom outlets.

In a critical industrial installation the electrical supply switchgear and distribution would be divided between two chambers which are not interconnected, with essential services duplicated. In the event of a fire or explosion, part of the plant and essential services would continue to function. This practice does not appear to be followed at spillway gate control stations.

The usual practice where central and automatic control is provided is to site local control systems close to gates. The latter are usually electro-mechanical controls. This satisfies one criterion of reliability of control systems, that there should be duplicate controls and that the two systems should be genetically different.

Earthquakes

The performance and safety of dams during earthquakes world-wide has been remarkably good (Charles *et al.*, 1991). Nevertheless the failure of a dam can have such serious consequences that earthquake safety evaluation of existing dams and of new constructions is a general requirement. Following an earthquake, the release of reservoir water can be a critical control function if the dam has been damaged by the seismic motion. Therefore, spillway gate installations and bottom outlets need to remain operational after an earthquake and should be included in the seismic analysis. As a New Zealand engineer expressed it, 'When the big one hits, the likely scenario is that massive power load will be dropped and spilling will quickly be necessary to prevent dam overtopping and serious damage to generating facilities' (Williams, 1996).

A review of incidents at dams that have been exposed to seismic events (Hinks and Gosschalk, 1993) shows that, while dam performance is, on the whole, fairly good, there are a number of events in which dams have suffered significant structural damage. In some of these cases the dam has subsequently failed entirely, although such failure has rarely occurred at the time of the earthquake; most failures have occurred either after a few hours or up to 24 hours after the earthquake.

Given this history, it is clear that directly after a dam has experienced a significant seismic event there is likely to be an urgent requirement for the water level in the reservoir to be lowered quickly both to reduce the pressure on the potentially damaged and weakened structure and to alleviate the consequences should the dam fail at a later time. Spillway gate and bottom outlets will be used for this purpose, with the spillway gates providing the greater initial capacity for level reduction.

Even in the UK, which most inhabitants do not consider an earthquake zone, there were some 201 earthquakes in 1998 in the British Isles and the surrounding continental shelf areas.

Seismic design of gates is complex, largely as a result of the earthquake inducing acceleration loadings acting in all three orthogonal directions simultaneously. This is particularly important in the sections of the gate which are subject to contact with the reservoir water. The overall effect of water adjacent to a steel element is to add to that element's mass and it is the product of the cumulative mass and the acceleration that generates the forces on the element.

Lateral movement of piers must be expected as a consequence of a seismic tremor. It is possible to design gates so that lateral movement of piers or abutments does not jam them by introducing collapse zones; however, this is rarely practised.

Cranes which operate turbine outlet gates are liable to be derailed due to an earthquake shock. Means to prevent crane wheels from jumping rails are established and are fitted where the risk exists. In military and naval defence applications, where electrical switchgear can be subject to severe shocks, shock absorbent mountings are installed. They do not appear to be used for electrical panels controlling gates in areas of high seismicity. An independent view is that the precautions necessary to ensure functioning of spillway gates and bottom outlet installations following a major seismic shock are mostly inadequate and that many appurtenances are at risk.

Damage and disablement of gates following an earthquake are not the only factors to be considered; blocking of access to an installation due to a landslip or damage to roads can inhibit emergency work.

Frequent operational problems or deficiencies

While reservoir appurtenances have a good operational record overall, there have been cases of failure, some potentially catastrophic, as well as areas where persistent operational problems occur. Specific causes of faults (Lagerholm, 1996) were:
- limit switch function;
- ice problems (in Northern Europe, Canada and some states of the USA; presumably this must apply also to Eastern Europe and other parts of the world subject to severe winter weather);
- seal leakage, which in winter can cause freezing of the gates;
- failure of heating systems;
- trunnion bearing problems (the most frequent source of faults) and
- loss of communication links.

To these must be added:
- gate vibration;
- cavitation at the bottom outlet of high head dams;
- silting of the intake to bottom outlets;
- lack of regular exercise of bottom outlet (also mentioned by Lagerholm, 1996);
- floating debris in extreme floods;
- electrical cable fracture and
- clogging of the intake and silting of water operated gates.

Of lesser frequency are:
- control system malfunction

- uncontrolled descent due to hoist brake failure (two known cases – one reported by Lagerholm, 1996)

There are also records of problems or breakdown of vertical lift spillway gates, bottom outlet rolling gates, pinrack operated gates, and others.

Risk assessment of gated structures of a reservoir

There is sufficient evidence of failures and malfunction of reservoir appurtenances for spillway and bottom outlet gates to be included in a risk assessment of a dam. Determination of reliability must include potential liability due to design, operation and maintenance, operator training, inspection and supervision and record-keeping of incidents.

Maintenance can be deficient, variable at different stations of the same authority, or completely absent because there is no maintenance budget or authority to order spares. The latter was the case at hydropower plants in a tropical country.

Methods of risk analysis

A number of detailed hazard and reliability assessments of barriers have been carried out. A barrier is constructed specifically to deal with the hazard of flooding, while the function of a reservoir may be electricity generation, water supply or flood storage and the hazard is perceived as a consequential risk. This may explain why, until recently, more detailed assessments of reliability were carried out on barriers than reservoir flood control structures.

There are a number of definitions of risk. The simplest one is 'The likelihood of occurrence of adverse consequences' (McCann et al., 1985). For the purpose of quantifying risk, the definition by BC Hydro (1993) is more useful: 'A measure of the probability and severity of an adverse effect to health, property, or the environment. Risk is estimated by the mathematical expectation of the consequences of an adverse event occurring (i.e. the product of 'probability × consequence').'

Risk analysis must by definition include probabilistic events, although they may sometimes be implicit. Risk assessment is a combination of art, judgement and science (in that order) constrained in a formalised process (Bivins, 1984).

The most detailed methods used in risk analysis are fault trees and event trees. Fault trees allow the diagrammatic presentation of components that may lead to failure in a system element. A general failure event – the event to be analysed – is at the top of the fault tree, the remainder of which is formed by specific events which can potentially lead to the failure. Analysis of the fault tree results in determination of minimal cut sets, which are the minimal combination of events which cannot be reduced in number and whose occurrence cause the top event. Calculation of the probability of occurrence for each minimal cut set is carried out from the probabilities of the basic events. A fault tree for failure of a spillway gate installation is shown in the Appendix.

An event tree represents all possible sequences of events which could result from a given initiating event. Unlike a fault tree, it works from the specific to the general. It therefore traces how failure sequences propagate. Branching is limited to 'yes' or 'no' at each system response. There are similarities with operational logic diagrams. An event tree for an earthquake which affects the dam is developed in the Appendix.

For analysing complex systems, computer programmes have been developed for the calculation of elaborate schematic structures. The one best known in the UK is AEA Technology's programme Fault Tree Manager. A previous version 'Orchard' was used in a reliability assessment of the Thames Tidal Defences and the barrages for the flood prevention of the City of Venice. Hoyland and Rausand (1994) discuss other programmes.

Other techniques which are used to identify different failure modes of appurtenant works consist of structured questions which help to analyse the system. Examples are failure modes and effects analysis (FMEA) and failure modes, effects and criticality analyses (FMECA). BS 5760: Part 5 (1991) describes these as methods of reliability analysis intended to identify failures which have consequences affecting the function-ing of a system within the limits of a given application, thus enabling priorities for action to be set. Hazard and operability study (HAZOP) is another analytical tool which concentrates on identifying deviations from design and operating conditions. These techniques use worksheets which are filled out during analysis of a system to document a qualitative assessment.

In dam engineering, a probabilistic risk analysis (PRA) is used as a basis for mak-ing decisions when selecting among different remedial actions, and to determine prior-ities. It is helpful when carrying out PRA if there is a collection of statistics, but this is not essential. When assessing gates and valves, an analysis of service records is a useful guide. This is also used when assigning failure probabilities to fault and event tree branches.

Reliability assessments based on fault trees of the Thames, Barking Creek and other storm surge barriers comprising the Thames Tidal Defences were carried out (UKAEA, 1987; Duke and Hounslow, 1990), and two reliability assessments of the design of the barrages for the flood defence of the City of Venice (AEA Technology, 1989; Lewin, 1993).

The Rykswaterstaat in the Netherlands has carried out similar assessments on bar-riers for flood protection of the Netherlands. Some results of a risk assessment of the New Waterway storm surge gate were given in the papers by Ieperen (1994) and Janssen et al.(1994). At spillway gate installations a fault tree reliability assessment was carried out as part of the deficiency investigations for the Seven Mile Dam in British Columbia (Klohn, 1996). This was a major undertaking – the documentation of the investigation of reliability for normal conditions is extensive, comprising three volumes.

In his paper, Lagerholm (1996) mentions that fault tree analysis has been per-formed in Sweden on different types of spillway gate functions. The wording suggests that these were not total system assessments.

The construction of fault and/or event trees and the production of minimal cut sets, together with the computational work required, involves considerable man hours. This type of analysis is considered necessary in special cases such as the Seven Mile Dam, where the operation and reliability of the spillway and drainage systems are crucial to the safety of the dam, or the Folsom Dam where collapse of a spillway gate has resulted in consideration of a fault tree assessment of the spillway system.

In most spillway systems there should be adequate redundancy, so that the malfunction of a gate does not result in a serious risk. If redundancy is provided the overall system reliability depends more on common cause failures, that is, on an event which affects the total installation, such as loss of power supply, a fire or an earthquake. However, redundancy of gates is rarely provided for an extreme event.

Even failure modes, effects and criticality analyses (FMECA) and hazard and operability studies (HAZOP) can involve much technical manpower. They are usually carried out by a team of engineers and technicians familiar with an installation, and can result in lengthy evaluation of specific elements of the control structures.

Where the operator of several dams requires an initial hazard assessment of a number of reservoir appurtenances of different design and age, methods of assessment based on the systematic application of engineering judgement are sometimes used. In Norway, a simplified risk analysis is being applied to dam safety. Scottish and Southern Energy uses a similar approach to determine priorities for maintenance and improvement of spillway gate and reservoir bottom outlet structures. The analyses could be more appropriately called 'systematic application of engineering judgement'.

Fault trees and minimal cut sets are important tools to assess the reliability of a total installation and for quantifying the contribution of sub-systems and major components to the failure of the top event. They are not, as a rule, extended to include details such as limit switches, an important vulnerable element of gate hoists, local leakages of seals which can cause gate vibration and freezing up of side seals at spillway gates in winter, and others. Unless data of operational problems over an extended period of time are available, it is difficult to assign failure probabilities to these and similar elements. This does not apply to the electrical supply and distribution systems of spillway gate and bottom outlets. General and detailed statistical information is available to assign a failure probability to each element and the result will more accurately reflect the failure probability than the parallel assessment of gates and their mechanical features.

Fault tree reliability assessments are a valuable tool to determine the overall integrity of an installation in relation to the risk of the dam and reservoir. The inclusion in the risk assessment of management, operator training, operational procedures, communication, possible malicious action and failure of advance warning systems results in a comprehensive assessment. In barriers, ship collision is an important risk factor and, more remotely, an aircraft crash.

Operationally, engineering assessments are required when the main objective is to determine the adequacy of maintenance, elimination or improvement of features or elements which are vulnerable, and the reduction in the probability of failures which can put a gate out of operation. A structured assessment system based on engineering judgement is probably the best means of achieving this.

A reasonable record of experience is available of design features of gates and valve systems which are likely to result in operational problems, or are indicators of risk. Instead of structured generalised questions which are the basis of HAZOP, more specific charts should be available for carrying out reliability assessments. They could take the form of diagrams and description of design features or the condition of a component and assign a number. The sum of the numbers would be an index of priorities and individual high numbers would draw attention to areas of urgent action. It would not form a probabilistic index, but if well constructed could be part of a risk assessment.

Reservoir control appurtenances are designed for extreme events and few have been subjected to exceptional loading. However, hydraulic conditions which cause gate vibration, while not necessarily extreme events, may not occur for years after commissioning. Hydraulic conditions, combined with structural, mechanical and electrical deterioration, can cause a risk and hazard because they are the coincident event of a number of probabilities. In a formal probabilistic investigation they may not show up, because it assumes that at each demand, the term used in reliability assessment for a gate movement, the structural, mechanical and electrical condition is assumed to be the same and that no deterioration has occurred. To factor wear and deterioration is difficult and quantifying it depends on judgment, subject to wide latitude.

Reliability indices

The probabilistic reliability derived from a fault tree analysis can be expressed as failure per demand. In the case of a spillway gate, this would be the opening of the gate.

For the Thames Barrier, this was 1.55×10^{-4} per gate per demand (UKAEA, 1987). Expressed differently, there is a chance that a single gate will fail to close on one on 560 closure demands, and that two of the ten gates will fail to close on one full closure in approximately 6000 closure demands.

In the hazard and reliability study of the flood prevention scheme for the City of Venice, failure was defined as flooding of Venice more than 280 mm above Venice datum. The design resulted in 1 event in 800 years (Lewin, 1996).

For the New Waterway storm surge barrier in the Netherlands, the derived reliability targets were:
- probability of not closing due to human or technical errors less than 10^{-3} on demand;
- probability of collapse less than 10^{-6} in any year and
- probability of not opening due to human or technical errors less than 10^{-4} on demand.

For the Seven Mile Dam in British Columbia the reliability analysis (Klohn, 1996) resulted in:
- probability of failure of spillway gates to open due to environmental hazards 9.68×10^{-6};
- probability of failure of spillway gates to open due to electrical or mechanical failures 2.07×10^{-7} and
- probability of power supply unavailability to the spillway gates 2.07×10^{-7}.

A good industrial system standard is one failure in 10^{-4} per demand. The reliability of a spillway gate installation depends on whether all the gates can pass the probable maximum flood (PMF) or the half PMF. The usual practice in a multi-gate spillway system is that a thousand year return period flood can be passed with one gate out of operation. A failure rate of 10^{-4} per gate per demand would appear to be an adequate assurance under these conditions. If the gates cannot pass the PMF a lower failure rate would be appropriate. This would depend on the hazard resulting from a gate failing to open under flood conditions. Some spillway gates, especially older ones, would not qualify for a failure rate of 10^{-4} per demand or a more severe criterion.

Bottom outlets consist more frequently of a single operating gate with a back-up gate or a discharge valve backed by a butterfly valve or a gate. Two or more parallel fluidways are less frequent. It is suggested that a failure rate of the order of 10^{-5} would be appropriate where only one gate with a back-up gate is provided. Whether a higher reliability is required than that of a spillway gate installation depends on the risk associated with failure of the bottom outlet.

Failure probability rating for electrical services, both for details and systems, are available and are statistically valid. This includes standby generating plant.

For spillway gates, their hoisting machinery and control systems, failure probabilities have to be assessed from service records, known incidents or structural and mechanical plant which have some similarity. The available data will probably be of low statistical validity. The selection of a failure probability for each item of a fault tree branch will therefore involve a significant element of engineering judgment. Such judgment, whether exercised by an individual or collectively, depends on experience.

The problems encountered with bottom outlets are more frequently the interaction of structural and mechanical aspects with hydrodynamics. Assignment of failure probabilities to fault tree branches when investigating bottom outlets is therefore even more dependent on judgment and knowledge of theory and practice.

When gate vibration is taken into account, the problem is compounded. ICOLD Bulletin No. 102 gives guidance on gate vibration, although it is difficult to relate it to practical problems. Other publications (Naudascher and Rockwell, 1994; Kolkman, 1979, 1984; Lewin, 1995) and many research papers are helpful but require knowledgeable interpretation.

A number of recent technical papers show that dam engineers are accumulating records of component and system failures of hydraulic equipment, and that these are being systematically analysed. For maximum results these should be carried out on a wide scale, and preferably on an international basis.

Some technical papers recorded failure events which resulted in serious hazards. This would not have been highlighted in a conventionally constructed fault tree and would have resulted in a low failure probability. Integrating experience and knowledge on a wide scale may identify areas where reliability and hazard resulting from a rare combination of factors would result in a different construction of a fault tree, or simply indicate the need for remedial action. A fundamental difficulty is to factor wear, deterioration and corrosion into any reliability assessment.

Conclusions

In moving hydraulic structures, gates or valves are the main reasons for failures. The design of control structures must therefore start with an understanding of the behaviour of the medium to be controlled, water.

Hydrodynamics is difficult to quantify but extensive research papers are available of model studies of prototype investigations of gate vibration. Design guidelines have been formulated. However, they require some understanding of hydrodynamics. In investigating actual cases of gate vibration, it is often disappointing to find that past mistakes have been repeated.

Hydraulic control structures are designed for extreme events and a long operating life. For reservoir gates it can be the probable half maximum flood which is assigned a 25,000 year return period. The earthquake return period and peak ground acceleration for risk assessment and design purposes depend on the classification of the dam and its location. The classification factors are the volume of the reservoir, the height of the dam, evaluation requirements if the dam sustains damage, and potential downstream damage (Charles *et al.*, 1991). Experience of extreme events is therefore rare and the worldwide collection of data is essential. For dams, this is carried out under the aegis of the International Commission of Large Dams which has its headquarters in Paris. The pooling of operating experience of reservoir control gates is less systematic than for dams and is very deficient for seismic events.

The structural, mechanical, electrical and control systems of gates and valves are subject to an operating life in excess of normal industrial plant. Some gates continue to function with equipment which is antiquated and cannot be replaced without significant redesign. They pose a special problem when assessing risk.

Risk assessments, especially at the design stage, can ensure a high standard of reliability of an installation. A large percentage of the total risk of failure on demand of a gate or valve is due to human factors and in many risk assessments they will predominate. These comprise organisational communication, training and maintenance deficiencies. The variability of the standard of maintenance of spillway gates and bottom outlets between different operating authorities, and sometimes within an organisation, is surprisingly common. It may be due to the isolation of dam installation and long distances between hydro plants, even if they are in common ownership. This devolves responsibility to local management and may explain differences in standards of maintenance.

Regular exercising of moving structures like gates and valves is essential. The reliability to operate on demand decreases with time since the last test. The reasons why it is not done vary from loss of water, opening a spillway gate or a bottom outlet to the safety precautions necessary to protect river users during a sudden release of water.

On raising bottom outlets, reservoir sediment can be discharged and this could be detrimental to fish life in the river downstream of the dam. In an extreme case, at the Barasona reservoir in Spain, the silt had completely blocked the outlet. This became a dangerous problem. When the hydraulic control structures of one hydro generating authority were inspected it was noted that no maintenance was carried out. On questioning local management it was stated that there was no maintenance budget. It had been forgotten when the plant was installed. Hopefully this was an isolated case. At another dam, the operators had never opened the bottom outlet although the reservoir had been impounded a number of years ago. The reason was that the operators were scared of the discharge.

How can failures be prevented or minimised, because entire prevention is not possible? Technically, by a better understanding of hydrodynamics and by a dissemination of operational problems and corresponding improvement of detail design and provision of redundancy where it is shown to be statistically necessary. Operationally, by recognising that the human factors are often predominant in an assessment of safety and reliability of reservoir, gate and bottom outlet installations. Two examples illustrate this. In the first case the dam and the spillway gates were located in a mountainous area remote from the nearest dwellings and the gates had to be manually operated. The access road could become flooded and the operators would have to reach the dam during the onset of a flood.

The second example was the effect of an earthquake where falling trees and a slip blocked the access road to the dam. The emergency generating plant, which was trailer mounted and towed by a Land Rover, could not reach the dam crest.

Recording of defects and minor operational failures is inadequate at many reservoir control installations. As a result, management and engineers responsible for a number of dams may be unaware of potential or even actual reliability problems.

Technical and organisational improvements depend on the identification of deficiencies. The importance and urgency of carrying out technical improvements in design and reliability is influenced by statistical data. The methodologies for identifying and quantifying hazard and reliability are a valuable tool in effecting improvements.

References

AEA Technology. 1989. *Reliability study for the Venice flood defences, Part 1: Reliability assessment of gate performance, Part 2: Hazard assessment.* Safety and Reliability Directorate, SRS/ASG/31466, Nov.

B.C. Hydro. 1993. *Guidelines for consequence-based dam safety evaluations and improvements* (Interim). B.C. Hydro, Burnaby, B.C., Canada.

Bivins, W. S. 1984. Risk analysis in dam safety programmes. *Proc. of the Conference Water for Resource Development*, Coeur d'Alene, Idaho, (ed. David L. Shreiber), ASCE, N.Y., pp. 115–19.

British Standards Institution. 1991. *BS 5760: Part 5. Guide to failure modes, effects and criticality analysis* (FMEA and FMECA).

Bureau of Reclamation. 1996. *Forensic report of spillway gate 3 failure, Folsom Dam*. Bureau of Reclamation, Mid-Pacific Regional Office, Sacramento, Cal, USA Nov.

Charles, J.A., Abbiss, C.P., Gosschalk, E.M. and Hinks, J.L. 1991. *An engineering guide to seismic risk to Dams in the United Kingdom*, Report, Building Research Establishment, Watford.

DIN. 1998. 19704–2: 1998–05 (Deutsche Normen). *Hydraulic steel structures, Criteria for design and calculation.*

Duke, A.J. and Hounslow, J. 1990. A practical application of reliability analysis to a working installation – the Thames Barrier. *I.Mech.E., Seminar Risk Analysis*, Nov, pp.13–33.

Hinks, J.L. and Gosschal, E.M. 1993. Dams and earthquakes – a review. *Dam Engineering*, **IV**, Issue 1.

Hoyland, A. and Rausand, M. 1994. *System reliability theory, models and statistical methods*. New York, John Wiley & Sons.

ICOLD. 1995. *Dam failures: a statistical analysis*. Bulletin 99, International Commission on Large Dams, Paris.

ICOLD. 1996. *Vibrations of hydraulic equipment for dams; review and recommendations*. Bulletin 102, International Commission on Large Dams, Paris.

Ieperen, A. van, 1994. Design of the new waterway storm surge barrier in the Netherlands. *Hydropower and Dams*, May, pp.66–72.

Ionescu, S. *et al.* 1994. Damage and remedial work during operation of several bottom outlets. *ICOLD, 16th Congress*, Durban, Q71, R7, pp.79–90.

Ishii, N. *et al.* 1977. Instability of elastically suspended tainter-gate system caused by surface waves on the reservoir of a dam. *Am. Soc. Mech, Eng., Fluids Eng. Division, Joint Applied Mechanics, Fluids Engineering and Bioengineering Conference*, New Haven, Conn, Jun., Paper No. 77-FE-25.

Ishii, N. *et al.* 1979. Dynamic instability of tainter gates. *19th Congress Int. Ass. of Hydraulic Research*, Karlsruhe, Paper C9.

Janssen, J. P. F. M. *et al.* 1994. The design and construction of the new waterway storm surge barrier in the Netherlands (technical and constructual implications). *ICOLD, 16th Congress*, Durban, C15, pp.877–900.

Kenn, M. J. and Garrod, A.D. 1981. Cavitation damage and the Tarbela Tunnel collapse of 1974. *Proc ICE*, Part 1, **70** February, pp 65–89.

Khan, K. A and Siddique, N. A. 1994. Malfunction of a spillway gate at Tarbela after 17 years of normal operation. *ICOLD, 16th Congress*, Durban, Q71, R27, pp. 411–28.

Klohn-Crippen Integ & Northwest Hydraulic Consultants. 1996. *Seven Mile Dam deficiency investigations; spillway and drainage systems, reliability for normal conditions for BC Hydro*. Task C8 – Part 1A, vol 1; vol 2 – Appendices A–L, vol 3 – Appendices M–0, Sep.

Kolkman, P. A. 1984. *Development of vibration free gate design*. Delft Hyraulics Laboratory, Publ. 219.

Kolkman, P..A. 1979. Vibration of hydraulic structures. In *Developments in Hydraulic Engineering – 2*, (ed.Novak, P.), Elsevier Applied Science Publishers.

Largerholm, S. 1996. Safety and reliability of spillway gates. *ICOLD Symposium, Repair and Upgrading of Dams*, Stockholm, June.

Lewin, J. 1993. *System reliability assessment of the definitive design of the Venice flood defences*. (Not published).

Lewin, J. 1995. *Hydraulic gates and valves in free surface flow and submerged outlets*. London, Thomas Telford.

Lewin, J. 1996. Mechanical aspects of water control structures. Dugald Clerk Lecture 1995, *Proc. Instn. Civ. Engrs Wat., Marit and Energy*, 118, Mar, pp. 29–38.

McCann, M.W. *et al.* 1985. *Preliminary safety evaluation of existing dams, volume 1*. Dept of Civil Eng., Stanford University, Vol 1, Report No.69.

Naudascher, E. and Rockwell, D. 1994. *Flow-induced vibrations – an engineering guide*. A.A. Balkema, Rotterdam.

Rajar, R. and Rryzanowski, A. 1994. Self-induced opening of spillway gates on the Mavcice Dam – Slovenia. *ICOLD, 16th Congress*, Durban Q71, R8, pp. 97–112.

Romeo, R. 1996. *Drawdown of the Barasona Reservoir*. Report by the author at the Symposium Repair and Upgrading of Dams, Stockholm, June. Reported in Hydropower and Dams, Issue Five, 1996, p.74.

UKAEA. 1987. *A reliability assessment of the Thames tidal defences. safety and reliability directorate.* SRS/ASG/31447. (Not published).

USNRC. 1983. *Safety of existing dams: evaluation and improvement*. U.S. National Research Council, Washington DC: National Academy Press.

Williams, I. S. 1996. After the earthquake – ensuring that the spillway can be operated when it really counts. *7th Hydro Power Engineering Exchange*, Hamilton, New Zealand, October.

Yano, K. 1968. *On the event of the gate destruction of the Washi-Dam*. Disaster Prevention Research Institute, Annals of Kyoto University in Japan, 11-B 1-17.

Appendix

The fault tree (Fig. A.1) investigates possible failures which contribute to cause the top event. In this case it is 'Spillway gate fails to open when required to discharge flood inflow to the reservoir'. In practice, a spillway gate installation will comprise a number of gates and failure of a single gate, or even several gates, would permit the discharge of a flood less than the probable maximum or the design flood. Since interest is in a complete failure of the spillway capacity, the analysis is likely to be dominated by events with the potential to affect all of a set of multiple gates. Random independent failures of multiple components are clearly possible but the probability of occurrence is likely to be of a much lower order than common cause events, that is, an event which affects the total installation; for instance failure of the mains supply. While multiple independent failures have not been shown in the fault tree, they would be included in a real analysis.

The increase in inflow to a reservoir during a flood varies depending on the geography of the catchment area feeding the river or rivers discharging into the reservoir. Snowmelt can usually be predicted well in advance, whereas a steep-sided river in a mountainous area would exhibit a very rapid increase in the rate of inflow, which could be as short as a few hours. A fault tree would usually omit any failure mode which can be rectified within the period between the onset of a flood and the time when multiple gates are required to open to maintain the reservoir level.

The impression of the illustrated fault tree might be that the events causing failure are fairly obvious. In practice, the fault tree would be developed further and the formal discipline of constructing it would ensure that all events are considered, not just structural, mechanical and electrical. The reliability of an installation depends as much on human factors, such as the actions of the operators, the organisation, the standard of maintenance, training and communication. External events, particularly when they can cause a common cause failure, will be included, such as a lightning strike, an earthquake, exceptional wave action due to a landslip into the reservoir or due to a storm of rare intensity.

The structure of the tree also allows for specific areas where redundancy provides higher reliability, or where combinations of failures could cause unexpected results.

The gates of the fault tree, which link the progression from lower to a higher event illustrate when a failure can be caused by either of several different occurrences, so called 'OR' gates, such as gates 3, 4, 5 and 20. The 'AND' gates 7 and 10 require that all the immediate lower events occur to cause failure at the next higher event.

When a fault tree has been fully developed it may raise a number of concerns which may result in design changes. Using data on the reliability of the components, the fault tree top events can be quantified. This provides a relative order of importance of the identified failure modes, as well as an estimate of the absolute failure probability. The most useful outcome is the insight gained into the potential vulnerabilities of the system. Even an approximate quantification can support concerns which are raised by good engineering judgement and experience.

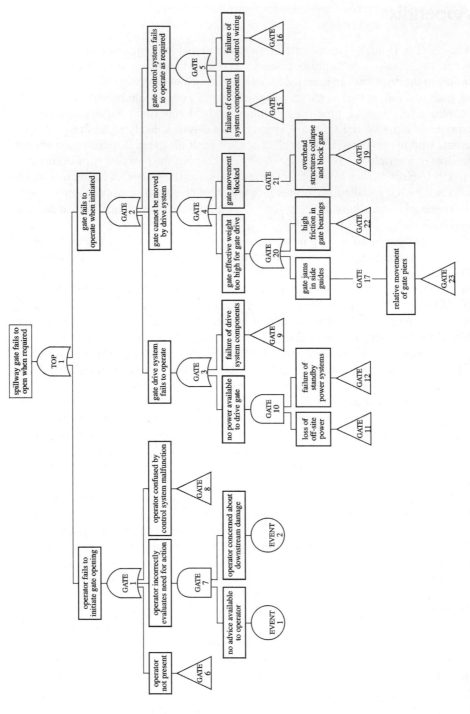

Figure A.1 *Fault tree for spillway gate failure (after Ballard and Lewin, 1998).*

The value of this type of analysis is to ensure that potentially important vulnerabilities are addressed at the design stage of a project or identified for action for improvement, and that a clear and auditable case is made for resolving each of the concerns. By the use of quantification, the issues can be prioritised and decisions can be made based on risk levels.

The following notes apply to specific headings on the event tree (Fig. A.2):

Damage to dam

During a seismic event of a particular intensity the dam may be damaged in different ways. The demand on the control systems will depend on the damage to the dam, or perhaps the extent to which this is apparent immediately after the earthquake. A family of event trees for different categories of dam damage may be used for a given earthquake. The analysis will need to reflect the operating procedures determined by the dam owners.

Failure of spillway gates

Depending on the extent of damage to the dam there will be a requirement for water drawdown over a specific period of time. For the present example it is assumed that this requires the operation of all the spillway capacity, initiated within one hour after the earthquake, followed by continuing operation of the bottom outlet. The full specification for the top event for the relevant fault tree will therefore be 'Failure of the system as designed to initiate full spillway flow within one hour of earthquake'. In practice this may always be the requirement for the more severe categories of seismic events because of the difficulty of determining the real extent of damage to the dam. However, the potential for downstream damage due to operating the spillway to its full extent will be an important factor in deciding the action recommended in the operating instructions.

Operator failure to recover

While the spillway gates may fail to respond as intended because of control or other failures, the operators may be able to recover the situation in time by various planned or ad hoc actions. The extent to which this is possible will depend on the time available but also on other factors such as whether the dam is normally manned, whether the operators are practised in fault finding and recovery, whether advice is available on a communication link which is still operating, etc.

Unintended operation of the bottom outlet

Either operator action or equipment malfunction may lead to spillway gate or bottom outlet opening when it is not required. The potential for downstream damage as a result of such opening requires that these events are considered in the analysis. For the purposes of the present example it will be assumed that the evidence suggests that operators will follow clear operation procedures and will not open spillway gates or bottom outlets unnecessarily. In the case of equipment malfunction, it will be similarly

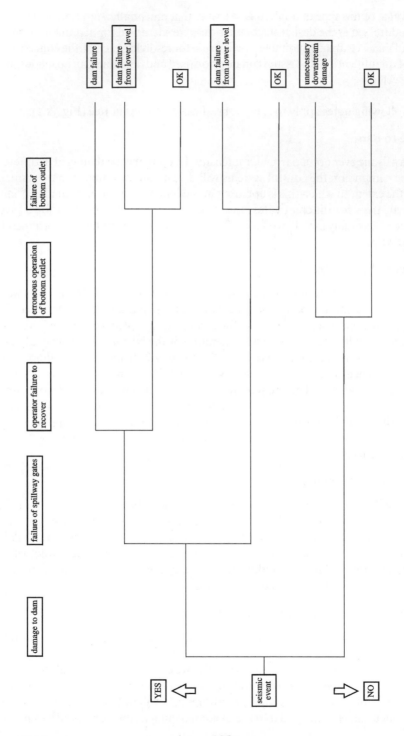

Figure A.2 *Event tree for seismic event on dam (after Ballard and Lewin, 1998).*

assumed that the operators would quickly recognise any unintended operation of the spillway gates because they are visible. It is less clear that they would either see and recognise control indication of bottom outlet initiation or that the discharge from the bottom outlet is visible. This event was therefore retained within the event tree.

Failure of bottom outlet

Dependent on the extent of damage to the dam, the bottom outlet may be required to open to provide continued lowering of the reservoir water level. Failure to open could lead to dam failure despite successful operation of the spillway gates, although the lowering of water levels resulting from that operation may serve to mitigate partially the consequences of a subsequent dam failure.

Event sequence consequences

The analysis is driven, and limited, by consideration of the events that concern the dam owners. This example is limited to those events with the potential to kill members of the public. In practice, the owner may have an interest in a wider range of consequences such as damage to generating capacity. Against this interest has to be balanced the greater complexity that would be required in the event trees and any subsequent fault trees which may be constructed.

Acknowledgements

Thanks are due to Thomas Telford for permission to reproduce material from Lewin (1995), and the following papers published in *The Prospect for Reservoirs in the 21st Century*, edited Paul Tedd, 1998, Proceedings 10th Conference BDS, Bangor, Sept. 1998.

> Lewin, J. Hazard and Reliability of hydraulic equipment for dams.
> Ballard, G. M. and Lewin, J. Should reservoir control systems and structures be designed to withstand the dynamic effects of earthquakes.

12 Diverse engineering failures

Alexander Kennaway

The thin-skinned tower

Narrative

In this example I represented the client who required a special-purpose, high tower and it was to be designed and supervised by a well-known consulting engineer. The proposition was to construct it by slip forming, a technique of which I had no personal experience. I checked with the best-known British expert who assured me that it was a well tried, reliable method. I sanctioned it and watched with interest as the work proceeded. After the concrete was properly set, the contractors cut holes to accommodate the supporting structure for the work room on top of the tower. They reported to the consulting engineers that the 'cover', i.e. the prescribed outer layers of concrete around the reinforcing steel, was only a fraction of that specified. Concrete allows the passage of water which will corrode the reinforcing steel when it reaches it. Iron oxide occupies a larger volume than steel, consequently it expands, causing cracks and outer layers ultimately to fall away thus exposing more steel. This significantly reduces the life of the structure.

The senior partner rang me suggesting that we had a problem and that we had to find a solution. I suggested that his grammar was faulty; he not we had a problem and it was not up to me to suggest a solution. I was after all only the client and did not want to share the duty of care owed us by the consultants and contractor. Nevertheless it was quite a facer. To blow up the tower and to start again would have put us way past the commissioning date required; quite apart from the cost and the shame. But I was known to have some experience of materials and hence he thought some input from me would be helpful. I consulted our solicitors who agreed my suggested approach; I offered to consider any solutions put to me and said that I would not object to the one with the least chance of failure, remembering that the expected life of such a structure ran into many decades. To my relief a suitable material was found with a written guarantee from a reputable organisation. It had to be trowelled on by hand all over the tower whose superficial area came to around 5000 m². It was done, but not at our expense.

– 194 –

What went wrong?

One section of the design office designed the reinforcing steel, whilst another designed the hydraulics to raise the slip form. Unfortunately neither consulted the other and no one oversaw both sections. Unfortunately, the hydraulic rams were set at the same diameter as the outer ring of reinforcement, thus pushing it a few critical centimetres further out and therefore reducing the radial thickness of the outer cover.

Lesson?

This is not the only occasion in my experience that senior engineers neglect the work of their juniors; perhaps they are occupied these days too much in meetings. Senior design engineers and senior draughtsmen used to go round the drawing boards twice a day when I was an apprentice and even when I was a design engineer. Perhaps modern management techniques and their priorities are at fault.

The floor that failed

Preamble

The architect of a major county council designed for them new, multi-storeyed offices of 10,000 m² superficial area; the exterior, landscaping and indeed most of the interior were admirable. Staff, of whom there were several hundred, began to move in with their office furniture and equipment. Before the building was opened to the general public it was very soon noticed that those parts of the flooring which served as walkways were collapsing. The floors were covered with a hard-wearing wall-to-wall carpet; beneath which were sheets of chipboard about 25 mm thick. These were separated from the reinforced concrete slabs at regular intervals of around 750 mm by long, rectangular battens about 150 mm high and 200 mm wide (Fig. 12.1). The arrangement was intended to provide a sprung floor and a space within it to house the cabling for electrical services and communications. Examination showed that the battens were made of a rigid foam of some synthetic polymer and that these had been crushed into fragments, even to powder and were cracked in many places. This flooring arrangement was a proprietary product promoted by the manufacturer, who acted as a sub-contractor to the main contractor of the building, who was directed by the county architect to specify this product.

The replacement of the whole floor area was plainly indicated and the cost would fall upon the county council which decided to sue both the main contractor and the sub-contractor. I was engaged as the technical expert by the main contractor.

Detailed examination

A short visual examination sufficed. The rigid plastic foam was almost certainly polyurethane, as I confirmed in later investigation. This material is suitable for insulating

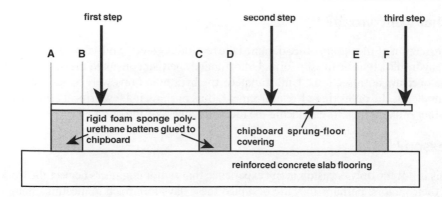

Figure 12.1 *Schematic diagram of floor structure.*

the walls of even large containers; it may be sawn, cut and in the correct densities is strong in compression but fails in almost any other mode of loading. Conventional sprung flooring is not dissimilar to the above; the battens are usually made of soft-wood and the top floor itself of a variety of timbers. Chipboard is relatively cheap but its use did not contribute to the failure of the flooring. The contractor had for his own reasons chosen to substitute this foam for the conventional timber. It seemed to me that he probably examined the compressive strength, found it to be more than suffi-cient and thought he had a structure fit for purpose and of merchantable quality.

Hypothetical cause of failure

The most likely cause was fatigue of the foam due to stress reversal brought about by forces imposed by walking on the floor. The initiation of a fatigue crack would be brought about by a small brittle fracture which proceeded through the material, re-ducing it first to lumps and then under further pounding, to powder.

The published sheets and handbooks giving material properties of these foams are silent on this matter; this is almost certainly because it did not occur to the manufac-turers that someone would attempt to use them in this mode. So there was no public warning to potential users ignorant of basic properties of such materials and indeed of the elements of fracture and fatigue processes.

Theoretical analysis

Figure 12.1 shows the floor and the changing forces upon it as a person walks across. At the first step, a downward force is exerted at that point and there are upward balancing forces at B and C. Because the chipboard is flexible and glued to the battens, there is also an upward force at A and D. As the steps progress, the batten are subjected to cyclic forces, rapidly leading to fatigue failure, small cracks at the surfaces, especially the top corners. The cracks develop, pieces fall off, stresses increase and the foam reduces to dust in places.

As someone walks on the floor across the line of battens, the stress in the interior vertical layers will vary from compressive to zero which also has a contributory factor to the fatigue process but is less damaging than the reversal from compressive stress to tension.

These foams are composed of lots of small holes produced by the evolution and expansion of a gas during the manufacturing process. The size of the holes and more important the dimensions of the surrounding relatively solid material forming a bridge over the holes are by no means uniform. It is easy to imagine that one or more such bridges are weaker than the rest and under a bending moment will crack and will do so rapidly if the stress due to that bending varies from positive to negative and is frequently repeated. The solid material is deliberately made to be rigid and relatively inflexible – unlike the sponges of the same materials used for washing, which contain large amounts of plasticiser for that purpose. Therefore it is very susceptible to brittle fracture. Such materials are well-known to fail easily in fatigue at stresses well below the average to cause failure in a single application. In this way the initial breakage proceeds through the body of the batten. Owing to the random nature of the micro-formation of the foam the crack proceeds randomly taking the path of the weakest material. This explains why the cracks and fractures were not the same in each batten investigated; a further, but less important, factor is that the walking patterns are themselves not identical.

Experimental supporting evidence

A rig was constructed to subject a sample of this foam to repeated reversal of stress at levels calculated to represent those exerted by a 50kg person walking across the floor with the dimensions shown in Fig 12.1. The material began to fail almost immediately; after a few hundred cycles cracks developed right through the batten, individual portions fell away and powder formed soon after.

Lessons

- No engineer or contractor in business can afford to be ignorant of basic material properties. Such understanding must be part of their basic education. These data are to be found in any decent textbook on strength of materials.
- It is not sufficient for such people to choose materials of constructions merely by looking at published material properties.
- It should not be necessary for the purveyors and manufacturers of materials to imagine all the possible ways of misusing their products and to overload their publications accordingly.
- Architects are especially at risk for many reasons. They cannot know everything about each and every material and the modes of their use in a complex structure, but they should have the humility to take the precaution of consulting competent engineers and materials practitioners about every aspect of their designs. Architectural education seems however to spend inadequate time on these matters and fails to engender sufficient respect for technology amongst their gradu-

ates who, contrary to public opinion, too often consider themselves to be positive contributors to public art.

• It is unwise to take on trust the claims of a 'novel' product especially if it forms part of a major structure where the risk of failure would lead to heavy loss of life, function or money. It would be of assistance if the insurers employed surveyors with the appropriate knowledge and understanding.

Marine containers that leaked like sieves

Preamble

A large number of standard containers 8ft × 8ft × 20 ft and 40 ft were commissioned by my client from a reputable contractor, to carry edible produce between the UK and the Antipodes. At the end of the first voyage it was found that the containers were partially filled with water; the contents had been ruined. I was engaged as the technical expert in the inevitable, ensuing litigation between my client, the contractor and the supplier of the materials forming the container. The containers could be carried on deck as well as in ships' holds.

Inspection showed that the container was formed from sections to which sheeting about 2 mm thick was riveted. Both were manufactured from glass-reinforced plastic (GRP). The doors, hinges and other fittings were of suitable metals. GRP sheeting and sections had been widely used for boats large and small ranging from dinghies to yachts and mine sweepers for the Royal Navy. Properly fabricated there was no reason why water should penetrate the sheeting, but this seemed the most likely means of the fairly considerable ingress of water.

The documentation showed that the contractors offered to build the containers, not from pressed and laminated or moulded sheets as was normal, but from sheets and sections formed by a process called pultrusion. This was a well-established technique for the production of certain elements but was a relatively new process for the formation of wide sheets. The advantage lay in the fact that wide sheets as well as sections could be produced to any required length, thus avoiding the hazards associated with joining fully 'cured' sheets.

The manufacturers using this process were well-established, part of a competent engineering group with excellent design engineering and R&D establishments. They assured the contractor that the material was in all respects of merchantable quality. With that background and pedigree the contractor accepted the assurances and in turn passed them on to my client. In evidence in court however it appeared that the manufacturers did not enquire concerning the final use to which their components were to be put and my client did not seek assurances that they were waterproof. In court the former sought to deny that the sheets were anything other than fully waterproof; their senior counsel suggested forcibly to me that my client should have enquired, to which my retort was that if that was the case it applied equally to the manufacturer who should have discovered the use to which the containers were to be put. This was especially valid since the normal duty of such containers is to spend long periods in the open. The Judge agreed with me.

In the pultrusion process, strands of glass fibre are pulled through a trough of liquid monomer, consolidated and then cured by heating the laminate until the monomer is polymerised. It exits from the final forming die in its final form, cut to length and stacked. In all the processes of forming a GRP section the main problem is to ensure that the fibres are fully wetted by the monomeric resin, leaving no spaces for monomeric gas or air. If this occurs, the section is not only mechanically weakened but the bubbles are liable to burst and form a crack to one or other of the surfaces, thus providing a leakage path for fluids whether liquid or gaseous. The water vapour permeability of a well consolidated section is discrete, small, known and published but is, in most applications, acceptable. Such a section will absorb and retain a small amount of moisture which may, depending on climatic conditions, also evaporate from the surfaces but will not pass significant amounts of water, whether salt or fresh. With hand lay up techniques a high standard is normally attained. Indeed the high quality achieved even in the early days by boat builders previously inexperienced in handling such materials is remarkable. Yachts may be damp but the inner surfaces of the GRP hull is not normally wet to the touch. However the process is slow and laborious; small wonder that people searched for continuous processes requiring very little labour.

Pultrusion has been used, successfully, to make thick sections of modest transverse dimensions, but significant length. These were advertised as being successful in many applications such as buildings where strong lightweight structural members were needed. By the nature of the application it is doubtful if any deficiency in resisting water transfer across the section would have been noticed. It came out in evidence that the manufacturers had never before supplied sheets for such a duty and that this sale was by nature of a proving experiment for them.

Preliminary investigation

A study of the process documentation together with the relevant drawings did nothing to assure me that the pultrusion process used by the manufacturers provided anything like a certainty that the fibres were fully wetted and that there were not going to be any gas or air bubbles in the sheet. I could therefore not dismiss my original suspicions that the sheets were the primary cause of ingress of a lot of water into the container. I therefore set about devising tests to establish the relevant facts.

Figure 12.2 shows a theoretical path through which water might penetrate a faulty sheet. If the outer gel coat of polymer is not free of a crack or small gas bubble, then water will enter. It may be that the crack or bubble extends right through to the inside of the sheet, but this is unlikely in practice. The more common mode is for the water to find a path along the surface of fibres that were not fully wetted by monomer and then to the inside through another hole or crack. The 'working' of the sheet, providing small but repeated movements will, of course, extend the nature of the cracks or gas bubbles. Microscopic examination showed plenty of such small faults, some continuous.

Confirmatory experiments

Sheets from the containers supplied together with other sheets direct from the produc-

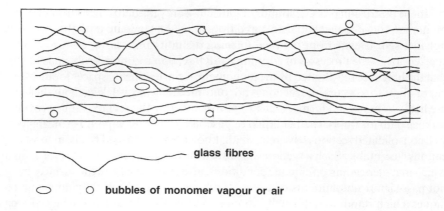

Figure 12.2 *Resin coated glass fibres in a sheet structure.*

tion line were laid horizontally on tables, with a space beneath them and their upper edges built up with sections glued to the periphery and water admitted to the resulting shallow tank to a depth of about 25 mm to simulate the pressure of rain or sea spray driven at speeds likely to be encountered at sea or on land. After a short while, sometimes not more than a few minutes, drops of water appeared on the previously dry underside of the sheets, providing supporting evidence of permeability of the sheets to water not merely to water vapour.

This evidence together with the refutation in court of evidence provided by the experts for the defendants was sufficient to satisfy the court who found in favour of my clients.

Lessons

- Extrapolation from one set of conditions to another is risky.
- It is a risky commercial practice to test a novel process and its products on a customer. This is especially true for an application new to the producer.
- Technical service and sales departments should resist such sales. Internal evidence found 'on discovery' suggested that they tried but were overruled by the production and technical departments.

Double glazing can become a tank for flat fish

Preamble

Much science but inadequate technology was applied a decade or two ago to the enhancement of glazing systems for large office blocks. One successful aspect is widely seen; the outer of the two glass elements in the window has a very thin layer of a gold-coloured metal applied to it under vacuum. This allows light to enter but reflects much of the heat radiated by the sun, thus reducing the load on internal air conditioning systems. Good experience encouraged the application to ever larger windows, some of truly large dimensions, reaching perhaps several metres in breadth and in height. In some of the office blocks in London, it was noticed that the space between the glass sheets contained water, occasionally to a considerable depth. Had the water contained microscopic fish of a flat variety, they might well have grown and provided some enter-tainment for the office workers. Since the source of the water was rain, this did not happen. These cases attracted much attention, from all concerned; some cases were litigated. They publicised the findings of the experts involved; I was one of them.

Preliminary examination

The double glazing was of a proprietary type, it was ingenious and had much success all over western Europe when restricted to small windows; it was the large ones that failed and then apparently only in England. The windows were constructed from the standard two sheets of glass, one of which, the outer, had the gold spattered on its inner surface. The space between the windows was created by soldering lead sections around the periphery to both inner surfaces, the space was evacuated and the whole sealed with strong tape or a mastic and inserted into the window space.

There were two issues, firstly how did water get onto the outside of the window and then how did it penetrate into the space within the double glazing.

The first issue was readily resolved by inspection of the drawings and the detail of the windows and their surrounds. In the cases where water could lie around the periph-ery of the double glazing unit, it was found that it could enter the surrounding space by several routes, but there was no provision for drainage of the surround.

When water had penetrated into the gap within the glazings it was noticed that the lead separators, especially on the long, vertical sides showed longitudinal cracks, usu-ally midway between the sides.

The problem was to provide a satisfactory explanation for these cracks, then to show that water did indeed enter through them and to demonstrate a mechanism that applied in the precise circumstances of failure and yet which was inoperable in other cases.

My hypothesis

I reasoned that the lead separators had been subjected to recurrent, longitudinal shear strain, that this occurred because the temperatures of the two glass panels were differ-ent, one being close that of the rooms and held static in the building whilst the other fluctuated depending on the amount of radiated heat falling on it when the sun was shining and when it was not. The lead failed due to fatigue in shear.

This idea depended on several subsidiary mechanisms, which we will take in turn.
- Why did it apply to large and not to small windows? Probably because the total movement is larger in a big window.

- Why did it apply to windows with lead separators and not to those made with other metals such as aluminium or steel? Perhaps because lead is subject to brittle fracture more quickly than other metals. However there were no published data in the handbooks at that time on fatigue properties of lead; probably because no one thought it would be used in such duties. This aspect had therefore to be tested by practical experiment. Furthermore, the use of the surface that reflected heat seemed to be confined to those manufacturers who combined this system with the use of lead spacers; I found none with spacers made of other metals but which used other metallic spacers.
- If direct radiated sunlight was responsible why was this not more prevalent for example in the Iberian Peninsula where the identical system was used for large windows in office blocks? After all, the number of diurnal changes in the life of the windows would be the same for Spain and for England, and the summer temperatures would be higher. The most plausible answer was the fact that in England in summer time, there was frequent cloud cover during the day, increasing by an order of magnitude the frequency of temperature change in the window.

Experimental support for this hypothesis

The temperatures of both panes of the windows were measured during several summer days. Thin wires were stuck to the window glasses and using a measuring microscope their relative movement both longitudinally and laterally was measured over long periods. This data was compared with times of direct sunlight and cloud cover and for night and day time. Using published data for the exact glass used in the windows the expansion due to thermal variation was calculated; very good agreement was obtained with the observed changes.

Several medium sized windows were made up by the manufacturer and subjected to the strain – not the total movement – calculated for the big windows; the periodicity was set at twenty minutes, approximately the same as that observed during that summer in London for sun and cloud cover. Several hundred reversals sufficed to induce longitudinal cracks of the type seen in practice. Water was allowed to lie on the periphery and penetration was observed.

Conclusions

- The experiments provided satisfactory confirmation of the basic hypothesis of failure. This was presented in court, subjected to the usual attempts at refutation but convinced the court.
- In my view the manufacturers of the system were unfortunate; their products were successful in smaller windows and everywhere in western Europe, even in countries with as much sunlight as in Spain. There was no published data on failure of lead through fatigue.
- However they had not thought through the possible mechanisms of failure described above. It would have saved them a lot of money had they done so but in normal commercial practice, such a depth of thinking through possible failure

mechanisms would have been considered by top management eager for commercial exploitation, as finicky and academic. After all the failure of windows in buildings is not the same as the failure of windows in a jet airliner, as in the first *Comet*. These days however, well educated engineers, physicists and chemists are more commonly employed in business and it would be a wise manufacturer and designer who thought of these things. In the long run, especially in these litigious days, it would be a saving of money and of management time if people routinely asked the 'what if....' questions. The only people to suffer would be the lawyers and technical experts!

• The manufacturer might well have relied upon architects employing their window system to have assumed that if water could conceivably get into the spaces around the window, then it would and therefore they would provide proper means of drainage of those surrounds. It was unfortunate that some architects failed to take this prudent precaution. Maybe they still do; it would be useful to know if such matters are drawn to the attention of students.

'It came to bits in me 'ands, Mum'

Preamble

A large international company with a British subsidiary was renowned for its sub-assemblies used in road transport. These products incorporated a lot of springs, which it bought from a well-tried supplier with a history of reliability. The overseas head office demanded that the British firm embark upon a process of reducing the price paid to suppliers. When the spring supplier refused, the next instruction was to search for a cheaper supplier. One was found which offered an extremely modest price reduction. The customer sent its quality assurance manager to vet the offer. He reported with many misgivings, stating that this firm was only to be trusted to supply products which were not to be used in a duty of prime importance. Nevertheless HQ insisted, the firm placed orders and delivery began according to a schedule that demanded tens of thousands to be delivered monthly. They were incorporated into the assemblies which were sent to the European customers. Complaints of breakages began to flood into the British firm, who returned to their original supplier and asked me to find out what was wrong.

Preliminary examination

The firm's laboratory was stacked with trays of the unused springs. Figure 12.3*a* shows a sketch of a very well known, conventional type of compression spring. I did what any engineer would have done, I picked one up by the ends and started to compress it between my fingers, almost as a doodle while I reflected. To my surprise it immediately snapped into two pieces. I looked more closely and found that the break was in two parts. The original was normal to the wire of the spring and NOT at the helical angle which would have occurred as a result of compressive loading, whereas the final fracture, due to my action was located helically (Fig. 12.3*b* and *c*). Examination with a low-powered magnifying glass showed that many of the springs supplied had this crack extending part-way across the wire. This kind of crack cannot occur as a result of loading the spring. These 'square' cracks, as I called them, occur as a result of brittle fracture in a high tensile steel, almost certainly due to an incorrect or delayed heat treatment after coiling.

Therefore there was something fundamentally wrong with the material properties of the steel which could have resulted only from a fault in the processes of manufacture. I had a suspicion but I wanted to confirm it. I consulted a colleague, an eminent engineering metallurgist. I showed him the broken pieces of several springs and told him the specification of the steel from which the springs were made. He immediately told me the trouble. This steel contained silicon and it required to be annealed immediately after heating and coiling, otherwise hydrogen formed during the process would embrittle the steel and it would break like a rotten carrot. This explained why 'it came to bits in me 'ands' before it was ever subjected to working loads. Hydrogen embrittlement of steels was well known to me in other applications, it is very destructive and if it occurs in the walls or components of pressure vessels will result in major accidents with risk of death and major destruction of property.

Subsequent steps

In response to a letter from my client's solicitors, the supplier denied the argument. But, the subsequent compulsory discovery of documents revealed the following:
  * The firm had received a standard circular letter from their trade association drawing attention to the need to anneal those steels within two or three hours of coiling. This letter was filed and there was no evidence that any notice had been taken of it.
  * The firm had produced a lengthy quality assurance manual, doubtless impres-

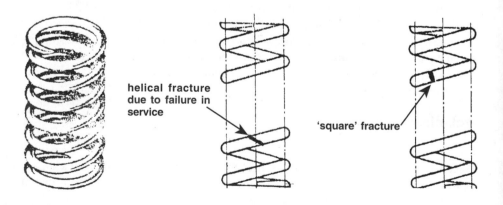

helical fracture
due to failure in
service

'square' fracture

Figure 12.3 *The helical compression spring (a) with expected failure mode (b) and the initial fracture seen in this example (c).*

sive to the uninitiated, but which contained no mention of the need for annealing rapidly nor did it specify the thermal conditions necessary to achieve the required results.

- On inspection, the thermal measuring instruments on the furnace were found to be reading incorrectly.
- The early batches that had been delivered and had proved to be faulty were coiled at the end of a week and only annealed at the beginning of the following week.
- A series of investigations were conducted on several hundred springs chosen at random from batches supplied by this contractor. Two thirds were faulty. Statistical analysis predicted that between 29.8% and 48.5% of the total population supplied by this contractor would have failed before being subjected to service loading.
- No springs supplied worldwide by other contractors failed in this or any other mode.

Conclusions

- People remote from the business should not issue instructions on detail. To overrule the local managers and especially to disregard the professional opinion of a competent quality assurance manager, however junior, is folly. This sort of management style is unfortunately increasingly likely when small firms form part of a bigger unit, which itself is bought up to form part of a multinational group whose operations are so disparate that the top management cannot possibly be competent to understand the nuances of each small business. If that trend continues then in all humility the top board must restrict itself to providing only general guidance and support services that they think the small subsidiaries cannot pay for themselves. In this respect they are often wrong; a healthy subsidiary can pay for professional advice locally and should be encouraged to seek it. There is plenty around. Are conglomerates in fact likely to be effective in every respect?
- Whereas cost reduction itself is a standard and useful aim of management, it has to be considered in the light of other attributes, especially performance, not to speak of maintaining a long-term, satisfactory relationship with suppliers. There is too much emphasis on money to the detriment of everything else these days.
- A quality assurance manual, however lengthy is no use if those who write it are ignorant of the essential elements that it should contain. Unfortunately we seem to be in an era where manuals with crisp, high sounding titles are seen not only to be the essence of good management but to take its place. Once it is on paper and the minions follow it like obedient soldiers, then plainly everything is satisfactory.

The incident puts one in mind of Disraeli's aphorism 'put not your trust in systems but in people'. Contemporary businesses would do well to follow this precept.

13 Reinforced aircrete slabs

Satish Desai

Introduction

Aerated concrete was developed in Sweden by an architect, Johan Axel Eriksson, at the beginning of the 20th century. He wanted a building material which could provide good thermal insulation with solid internal structure and low density. Timber could have these attributes but it would suffer from decay and require special measures to achieve fire safety. These developments led to what became known as autoclaved aerated concrete (AAC). One of its patented names is Ytong, which is made up of letters from two words; Y from Yxhult, the place where it was developed, and the remaining letters from betong, the Swedish word for concrete. The material has been the subject of many international symposia, including those organised by RILEM, an international association of laboratories engaged in testing and in research on materials and structures. RILEM Technical Committees have prepared a comprehensive guide on AAC, with information on properties, testing and design (RILEM, 1993).

AAC is used in two main forms of structural elements; lightweight wall units and reinforced or RAAC roof slabs and floor slabs. AAC, or aircrete, wall units have a wide range of use, for example in blocks. Popular applications of these blocks include construction of inner leaves of masonry walls, solid internal party walls and partition walls. Aircrete blocks have been found to be quite suitable as load-bearing elements in their protected environment, to carry the requisite axial compressive loads and to provide the thermal insulation expected from the wall construction. The other potential uses of aircrete blocks may include outer leaves of masonry walls, external walls and walls below ground level, where adequate care is essential to ensure their durability and protection from effects of the environment.

In the UK, serious serviceability problems were revealed during inspection of some pre-1980 roofs constructed with RAAC slabs. These problems were perceived as arising from incorrect installation and, perhaps, lack of appreciation of the behaviour of AAC being different to that of normal concrete. As a consequence, the UK Association of Autoclaved Aerated Concrete Products Association (AACPA) decided to call the slabs reinforced aircrete slabs to remove any misunderstanding caused by the reference to concrete in the original name of the material. This chapter concerns the issues

– 206 –

associated with installation of reinforced aircrete slabs and potential solutions to prevent recurrence of any problems in the future.

Characteristics of the material

Manufacturing process

Aircrete slabs are made by introducing gas bubbles into a cement paste thus reducing the weight and improving thermal insulation properties of the product. The principal ingredients are usually:

- calcareous raw material (normally quicklime or quicklime combined with Portland cement);
- siliceous raw material – generally silica flour or finely ground siliceous sand for example, quartz sand with SiO_2 content of 70% in most cases. Pulverised fuel ash (PFA) is used as an important source of siliceous material and not specifically as a binder component, and
- aluminium powder to react with the lime to produce hydrogen and provide the material with a large number of air pores plus a closed cell structure of calcium silicate hydrate.

The calcareous and siliceous ingredients are mixed together with water to form a low viscosity slurry. The mix receives a predetermined dose of finely divided aluminium powder just before it is poured into a pre-oiled large steel mould. The mould is of a size and dimensions suitable for the volume required to make a specific number of slabs. Reinforcement is made up in the form of cages, with longitudinal bars spot-welded to the U-shaped links. These cages are treated with protective coating (a cement-rubber latex coating was used in the past but manufacturers appear to have changed this practice and now use a bituminous coating). The reinforcement is so placed that it can fit precisely within the individual slabs, with an allowance for their subsequent trimming.

Initially, the mould is kept in a warm and humid environment and the slurry is left to hydrate for two hours. The aluminium powder soon reacts with the alkaline components and hydrogen gas is released, forming numerous bubbles in the slurry and increasing its volume by up to 220%. During the first 15–25 minutes, chemical reactions between the cement and water yield gels of calcium silicate hydrates and calcium aluminate hydrates, and slaking of quicklime produces gelatinous calcium hydroxide. This process converts the mix into a mass with stable cellular structure. At the end of this stage, the hydrogen gas escapes and it is replaced by air, and the temperature of the mix rises to about 75° or 85° C, making it stiff like a cake.

The next stage starts immediately, wire-cutting the cake to produce slabs. The slabs are profiled to shape at the edges and trimmed as appropriate to provide accurate plan dimensions, thickness and cover to the reinforcement. This process is helped by tilting the mould through 90° to allow such precision work of cutting and profiling on all sides. After cutting, the cake moves to the steam-curing process at 190^0 C for 10–12 hours. The steam pressure is increased at a specified rate and held at a level typically 11 bar absolute for a specified time, followed by controlled depressurisation of steam. This process is known as autoclaving, which promotes further reactions in the mix to produce crystalline hydrates and to bind together the intercellular matrix of aircrete. This provides the material with the strength required to meet the specification. Autoclaving provides additional benefits, for example:

- chemical and physical stability of the product, to improve its durability (e.g. resistance to sulphate attack and resistance to freeze-thaw); and
- elimination of efflorescence, as there is no lime left to be leached out.

Manufacturers have developed the process of making aircrete slabs with emphasis on control of materials quality and dimensional accuracy. They have also kept pace with environmental considerations, to ensure that:

- the material is now produced without pollutants and any hazardous waste;
- the process uses energy-saving measures through steam-curing at relatively low temperatures and recovering thermal energy for maximum efficiency; and
- wastage of raw materials is avoided and production trimmings are fully recycled.

Inherent properties of the material

Aircrete has properties and an internal structure different to those of normal concrete. Some major and specific characteristics of aircrete are observed as follows:

- Compressive strength is low (2.5–7 N/mm^2) and the ratio of tensile strength to compressive strength is higher than that for normal concrete. The direct tensile strength of aircrete is 15–35% of the compressive strength (RILEM, 1993). Typical flexural tensile strength is 22% of the compressive strength. These values are normally related to a stable gravimetric moisture content in aircrete, less than about 5%. Owing to its internal structure with pores, aircrete tends to suffer from reduction in strength of some 15% with an increase in moisture content from 5% to 10%. Aircrete with moisture content of 50% could retain only about 80% of its original strength.
- Compared with normal concrete, aircrete has a limited post-elastic stage, mainly attributable to its internal structure and reduction in ductility resulting from the autoclaving process. (Elastic range for normal concrete could extend up to a compressive stress of 30% of its characteristic compressive strength (f_{cu}). In non-linear finite element analyses, it is a common practice to assume a limit of 5 N/mm^2 for the elastic behaviour of concrete represented by the straight line part of the stress-strain curve. The non-linear part of the curve rises to the stress level $0.67f_{cu}$ and it is continued up to the limiting strain of 0.0035.)
- Modulus of elasticity (E_c) is low, corresponding to the low characteristic compressive strength and the limiting strain of 0.003. For example, E_c is 1.33 kN/mm^2 for aircrete with f_{cu} of 4 N/mm^2, compared with 20 kN/mm^2, modulus of elasticity of

normal concrete. E_c depends on the ambient relative humidity (RH). At an ambient RH of 100%, E_c could be some 88% of its value at the ambient RH of 50%.

- Aircrete has no coarse aggregates to provide any significant aggregate interlock. For normal concrete members, aggregate interlock contributes to their shear resistance and response to loads after micro-cracking has taken place.
- 'Raising' of the mix may influence the internal structure of reinforced aircrete slabs. It could be less dense for a short length above a bar obstructing the upward movement of the mix. Such 'shadow effects' may weaken the bond between aircrete and the steel.
- Owing to its porosity, aircrete is less resistant to moisture penetration than normal concrete. As such, most free alkalis in aircrete are combined during the process of autoclaving. The steel has to be provided with protective coating, therefore, since the aircrete in the cover region does not provide any corrosion protecton to the steel (RILEM, 1993).
- Some samples recovered from test slabs showed that the protective bituminous coating had come apart from the steel and adhered to the surrounding aircrete. The bond between the coating and the steel, therefore, is very important and it should be such that the steel remains effectively bonded to the aircrete.
- The moisture content in aircrete shortly after manufacture may be 25–35% by mass of the dry material (RILEM, 1993). It could soon dry out to leave some 10% of moisture. The manufacturers' advice is to delay installation of the products in building and allow the moisture content to settle at an equilibrium level below 5%. In roof construction, however, the builders tend to apply the bituminous roofing membranes on the top and paint on the soffit of slabs as soon as possible, which may trap the moisture and adversely affect the modulus of elasticity.

In the light of these special characteristics of aircrete, it would seem quite inappropriate that aircrete slabs should be treated as if they were low strength concrete members. To start with, the cube strength of aircrete is far below the lowest acceptable in reinforced concrete according to BS8110 (1997), i.e. 15 N/mm² for lightweight aggregate concrete. This minimum strength is essential for the material to have composite action with the reinforcement implied in the design rules and to give it protection from the environment, as appropriate for its intended use in the building. However, the earlier impression of its similarity with concrete may have misled the users and designers, resulting in some serious serviceability problems. This situation may have become worse because of inadequate installation and maintenance of roofs. Some such instances are briefly described in the next section.

Case studies

The Building Research Establishment (BRE) information paper IP 10/96 (1996) gives a detailed account of serviceability problems in pre-1980 roofs constructed with aircrete slabs. The main causes for concern were excessive deflections, ponding, cracks in the soffit and corrosion of reinforcement.

Figure 13.1   *Ponding on a pre-1980 aircrete roof slab.*

However, roofs with aircrete slabs which have performed satisfactorily are known to the author. The span-to-depth ratio for these aircrete slabs was as low as 14. Reports on these buildings show that they are inspected regularly, with adequate attention to the condition of the roof felt. Similarly, there are examples of problem-free floors in internal dry environments, built with bonded structural screed and continuity over supports. Such measures were apparently not available to installations described in this section, which suffered from problems similar to those described in the BRE information paper.

Some earlier installations have been demolished during the last 15 years, because of serious serviceability problems and unacceptable performance of aircrete slabs. Generally, the problems appeared to be attributable to inadequate understanding of properties of the material and lack of care in installation and maintenance of the roofs. The slabs are not made in the UK at present but it is possible to import them from other European countries. It is essential, therefore, to plan fresh guidance for the future, based on a study of some typical cases.

Case study 1

Aircrete slabs were used in a low-pitched roof with felt, for a fairly large steel-framed portal shed building in Somerset, constructed in early 1970s. Slabs spanned between the rafters at 4500 mm centres. The depth was 175 mm, giving a span-to-depth ratio of

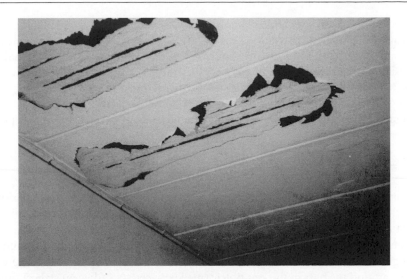

Figure 13.2 *Cracks and spalling on the soffit of pre-1980 aircrete slabs.*

Figure 13.3 *Corrosion of steel in pre-1980 aircrete slabs.*

26. The reporter has described this choice of depth of slab as 'on the limit of the manufacturers' recommendation', 25 years after the incident.

Unfortunately, two persons walking on the slabs during erection caused collapse of a few slabs and they were injured, suffering from a fall of about 10 metres. The shattered slabs showed very limited amounts of tension steel, four bars of diameter less than 6 mm, with welded cross-bars at the ends. These bars were observed to be stopped at 200 mm from the ends for the broken slabs. For most other slabs, longitudinal bars were visible and unprotected at the ends, which should not have been the case. Ironically, this was taken as evidence of tensile steel being present and available right up to the support and the slabs without such evidence (30%) were rejected and replaced.

Incorrect placement of tension steel and inadequate depth (high span-to-depth ratios) were cited as the major criticisms on this installation.

Case study 2

This study was carried out in 1976 on residential buildings in a suburb of London. It revealed a large variation in cracking and deformation of aircrete slabs, even when the properties of aircrete samples taken from the slabs and the design provisions were generally as expected. One of the main reasons for cracking was identified as failure of the longitudinal joints between slabs. These joints had been filled with fine aggregate concrete around the bars in the key-way formed between two adjacent slabs. Longitudinal cracks may also have been the result of an insufficient bond between the steel and aircrete resulting from the shadow effect described earlier.

Cracking may also have been the result of irreversible moisture movement in aircrete. Such moisture movement may have been brought about by fluctuating air humidity conditions in the dwellings. It is believed that each cycle may have left the slabs slightly shorter than before. The reason for such air humidity conditions was identified as the application of either excessive heating in the dwellings or its irregular pattern. Another effect of moisture on aircrete has been identified as reduction in its modulus of elasticity (E value). Tests on some samples showed abnormally low E values.

Case study 3

This case study was undertaken in 1981 to investigate problems in roof construction of a complex near London. The units were 4600 mm long and 610 × 152 mm in cross-section. The tension reinforcement was 0.4% and the cross-bars were 6 mm bars at 1000 mm centres. These cross-bars were apparently too few to meet the standards expected according to the modern practice (as described later in this chapter), although they may have been acceptable according to the practice prevalent at that time.

Tests on slabs showed that moisture content was high (>7%). In the absence of any protective environment in the surrounding aircrete, the steel had rusted. Corrosion had resulted in reduction of bar diameters to 9.5 mm in some places.

Higher span-to-depth ratios (30) may have been the reason for excessive deflections in the slabs. The situation may have become worse because of two factors; reduction in strength of aircrete due to high moisture content and reduction in effectiveness of tension reinforcement due to inadequate cross-bars. As a consequence, the roof membrane may have stretched at the supports, since the slabs were placed parallel to the fall or drainage. As a result, the designed slope of the roof would have been less effective, causing ponding at the low spots. Further aggravation in the condition of the roof was the result of temperature increase in the membrane at dry edges near the support, owing to the solar gain effect on dark materials. This would subject the membrane to reversal of thermal stresses at the edge of ponding, which would enlarge after the rainfall and then recede due to evaporation. Growth of moss and collection of dirt meant formation of dam-like barriers to water, making bad conditions progressively worse, leading to the failure of effectiveness of the protective membrane.

Inspection of soffits of roof slabs was made difficult by restrictions to access in many areas of this establishment. The main reasons were provision of false ceilings, services under the soffit and equipment on the floor. Dampness was visible in some places, suggesting that the slabs had cracks at the soffit. In the end, roof slabs were removed and replaced for a large part of the complex. The new work included increased falls in the roof and the gutters.

Case study 4

The BRE carried out inspection and tests on aircrete roof planks in 1991. They were also asked in 1994, to inspect two school buildings, where extensive deflections and cracking were reported. Their main findings were common to both cases. The investigation concluded that:

- excessive deflection and transverse cracking at the soffit were caused by long term slippage between aircrete and steel, resulting in loss of bond;
- corrosion protection to the reinforcement was breaking down and localised corrosion had started;
- however, the maximum load-carrying capacity of the planks was not significantly affected and they retained a satisfactory factor of safety.

With the transfer of ownership of some school buildings from the public sector to the governing bodies, the Department for Education (DfE, as it was at that time) wished to encourage proper surveys of such buildings, based on some authoritative guidance. The BRE Information Paper IP 10/96 (1996) provided such guidance, aimed at identification of aircrete slabs and effective resolution of problems, if any. It clearly stated that it does not apply to 'beam and block floors and walls which contain autoclaved aerated concrete (AAC) blocks in their construction or any other AAC components'. The paper also confirmed that there was no evidence so far to suggest that RAAC planks pose any safety hazard to building users.

Unreported cases

The cases described above probably represent a fraction of the number of inadequate installations of pre-1980 aircrete slabs in roof construction. It would seem that, sometimes, suppliers were called in and remedial solutions or removal of slabs were worked out without any formal record or feedback to the engineering profession, for reasons of legal liabilities and commercial interests. In Europe, similar practice may have prevailed. The author has been unable to obtain any report on European experience beyond some verbal responses to informal enquiries. These responses amount to instances of removal of parts of roofs, as tasks included in routine maintenance procedures, when problems were revealed during inspections.

Actions concerning buildings with pre-1980 aircrete slabs

Identification of aircrete slabs

Aircrete slabs can be identified on the basis of some observations:
- closed cellular structure and absence of coarse aggregate;
- low density and lightweight material (a broken piece could float on water);
- slabs with a width of about 600 mm and with a groove at the edges for forming joints between adjacent slabs (as shown in the sketch in Fig. 13.7).

Inspection and maintenance

Distress in most buildings could be perceptible to the users, for example, ponding on the roof and cracks on the soffit of slabs. However, a competent person should be called in to ascertain the extent of some critical factors; for example, deflections, corrosion of steel and bearing of slabs on the supporting members. Remedial measures to the slabs could be impractical and replacement of units and protective membrane may be the right way forward for slabs with deflections greater than span/100. Frequency of inspection should depend on condition of the roof and the nature and extent of critical factors causing the distress.

For roofs in better condition, regular inspection and maintenance should ensure that the protective membrane remains in good condition and that debris and dirt are cleared from the roof area including gutters. In some cases, it may be advisable to remove chippings and other items to reduce the permanent load and replace them with other suitable and lightweight materials. For example, a reflective membrane could replace chippings and reduce the load imposed by own weight of chippings and the weight of water retained.

In some cases, aircrete slab roofs have increased the problems generally associated with flat roof construction. Drainage outlets have been left marginally above the top surface of a structural slab, with the belief that a damp condition in the chippings could protect the bitumen-based membrane from solar heat. Such roofs have been most severely affected, with increased deflections and worsening of ponding. Remedial measures for such roofs should be based on lowering the drainage outlets and removing chippings as described above.

Safety and serviceability of future installations

In the UK, production of aircrete slabs ceased but the industry continued to do well with aircrete masonry blocks. This did not encourage seeking real solutions to the problems of the past. More important, it did not promote correct and effective use of aircrete slabs as a product well able to serve the interests of energy conservation in buildings.

In early 1995, the Verulam Column of *The Structural Engineer* included correspondence based on inspection of a school building roof. The correspondent expressed doubts about the suitability of aircrete as a structural material concerning anchorage to the steel and low strength of the material. It was proposed that the coating would be a potential 'bond-breaker'. The correspondent disagreed with the procedure (BS8110, 1997) which appeared to give aircrete or AAC a status alongside structural concrete. This article may have been the beginning of the efforts to take positive actions for the future, to consider use of aircrete slabs on the basis of correct understanding of the material's properties. This seemed very important from the building control point of view, as the slabs could be imported in the UK from other European countries with the advent of European Product Standards and removal of trade barriers.

Over the next two years, the problem concerning AAC was followed up by the Standing Committee on Structural Safety (SCOSS) and a letter was sent to a member of parliament. During this period, the author had taken some steps to convince the British Standard Institution Committee responsible for BS8110 that the standard should not contain the section dealing with AAC. This was achieved and it was very helpful in reassuring both SCOSS and others that an action was being taken by the Department of the Environment, Transport and the Regions (DETR). DETR followed a policy in responding to the queries on the following lines:

- Problems of pre-1980 RAAC slabs concerned mainly the serviceability and not the safety of the buildings as explained in the BRE IP 10/96.
- Such problems could arise mainly from lack of understanding of the limitations of autoclaved aerated concrete as a structural material.
- It would be inappropriate to rule out the use of RAAC slabs altogether, under all circumstances, and without any regard to the fact that there are instances of their successful installations in the UK and in Europe.

DETR preferred an impartial study, without any preconceptions, to raise awareness of the industry and clients on various issues related to aircrete as a structural material. It would have been inappropriate to take any steps which could be misunderstood and give incorrect impression to the users about suitability of all aircrete units as construction products. This was specially significant for the UK industry, since there were no reported instances of significant problems with internal walls constructed with aircrete units or blocks. It was essential, therefore, that the study should be based on factual feedback on some installations of aircrete slabs and laboratory tests on aircrete for studying its structural properties. Tests on aircrete would lead to the criteria for designing aircrete slabs, with AAC and steel reinforcement acting together, and an understanding of behaviour of aircrete slabs under various exposure conditions. This study would lead to a draft guidance document and it could be published to accommodate comments from all interested parties.

As a first step, a notice was published in May 97 in Journals of the Institution of Structural Engineers and Institute of Building Control, inviting information from those who may have experience in design and installation of aircrete slabs. The notice received a positive response and feedback. The correspondence was analysed and further tests were carried out to meet some suggestions. A brief account of some important aspects of this exercise are given in the following sections.

Research and study

General brief for research

It was essential to derive fresh rules for design of aircrete slabs, accounting for the realistic structural properties and behaviour of aircrete. It is understood that the earlier difficulties experienced in aircrete slabs may not be overcome by improvements in only the structural design. Comprehensive guidance including detailing and installation procedures is being prepared separately, with the support of aircrete manufacturers.

The guidance document will account for issues brought out by the BRE and the response received from engineers to the DETR notice in technical journals. The following points represent the general brief for the guidance document:

- Serviceability problems are more generally evident in roof installations than in floors. Floors are not usually exposed to a corrosive environment (such as moisture ingress), with the exception of floors in swimming pools or large kitchens. Moreover, they are invariably have a screed to protect the aircrete surface from abrasion. The screed has reinforcement, which provides continuity to the slab over supports and reduces mid-span deflection and cracking at the soffit. The screed also provides additional robustness to the floor and the building.
- Estimates of deflections should be based on realistic properties of aircrete. Depths of slabs in roofs and in floors should be adequate to limit the deflections. Such limitation should allow for any probable increase due to creep and potential effects of moisture. The limit may have to be more stringent in roofs, where such conditions are more likely and may result in ponding and damage to the protective membranes.
- Welded cross-bars are vital for the anchorage of tension steel. The study should investigate their influence on the behaviour of aircrete slabs.

Objectives for the guidance on design

A designer would not normally be expected to evaluate safe loads and deflections of aircrete slabs. Manufacturers normally supply the slabs to meet the specifications of loading, for the roof or for the floor construction. However, a design method is proposed to illustrate the case of simply supported slabs subjected to uniformly distributed load, supported by the limited test data available to the author. The manufacturers may wish to undertake more detailed research and studies and consider reviewing their technical literature. Such initiatives could also help in finalising the BSI draft European Standard prEN12602 (1996) on the subject of prefabricated reinforced components of AAC.

Structural design procedure

Test programme

Tests were carried out at the University of Leeds, mainly to study the post-elastic response of aircrete and the parameters required for estimating deflections. The tests used slab specimens 600 × 200 mm in size, loaded at two locations at third points on a span of 2.5 metres (Fig. 13.4). Cover to 8 mm dia. bars at the bottom was 20 mm, giving the effective depth as 176 mm. Three types of slabs were used with different amounts of tension steel (A_{st}): 201 mm² (4 nos), 302 mm² (6 nos.) and 402 mm² (8 nos.). Cross steel was in the form of 6 mm dia. U-shaped links at 250 centres, spot-welded to the bottom steel and to 2 nos. 8 mm bars at the top ends of the upturned legs. These bars are useful for effective handling of the steel cage. They should not be considered as effective compression steel, when they are placed far apart, only at the corners, and do not have any restraint provided by legs of links around them.

Characteristic strength (f_y) and modulus of elasticity (E_{st}) of tension steel were 500 N/mm² and 200 kN/mm² respectively. Measured compressive cube strength of aircrete was 3.8 N/mm².

Modulus of elasticity of aircrete (E_c) was derived to be 1267 N/mm², based on the strain corresponding to the compressive strength (f_{cu}) as 0.003. This also gave the Modular ratio 'm' (E_{st}/E_c) as 158.

Observations of test specimens

In general, all slabs developed predominant cracks leading to a sudden failure, nearly at the same location with respect to the applied load (Fig. 13.5). The mode of failure was very similar to that observed by the author during manufacturers' quality control tests on aircrete slabs. During the tests, the specimens did not have any visible flexural cracks. There was no evidence of yielding of tension steel or excessive compressive stresses. The crack leading to failure was nearly vertical at the tension face close to the

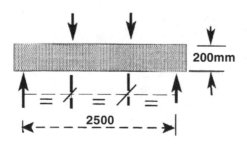

Figure 13.4 *Schematic loading (University of Leeds tests).*

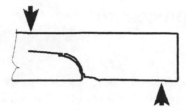

Figure 13.5 *Typical crack at failure.*

load point and curving up to the neutral axis level in the direction of mid-span.

These observations suggest that the failure could have been triggered by excessive tensile stresses in the neutral axis region, at the location of critical shear and bending moment under the applied load. The predominant crack could have widened with increasing applied load, leading to failure of anchorage to the tension steel. The slabs may have undergone the following stages (Fig. 13.6):

- With the tensile stress in aircrete at the soffit below its limit, a triangular stress block achieves equilibrium between tension and compression on either side of the neutral axis.
- With further weakening of aircrete in the tension zone beyond Stage I, a crack develops due to combination of shear and flexural stresses below the neutral axis at the critical location. (In case of the test specimens, under the applied load.)
- With a further increase in the applied load, anchorage force in the tension steel (F_{RA}) is resisted by the bearing pressure exerted on aircrete by the cross-bars, which maintains equilibrium with the compression block. This stage illustrates an important difference between normal concrete and aircrete. In reinforced concrete units, anchorage of tension steel is provided by its bond with concrete itself but the aircrete slabs depend on the cross-bars for anchorage of the tension steel.

It is proposed to adopt procedures for ultimate stage design and serviceablity design (deflection check) based on the test results.

Ultimate stage design

The following steps are proposed for evaluating the various parameters as shown below:

- Tension in steel depending on anchorage provided by cross-bars (F_{RA}).
- Neutral axis distance from the compression face (x), to provide equilibrium between F_{RA} and compression in aircrete given by the limiting compressive stress of $0.67f_{cu}$.
- Moment of resistance of the section (M_u) based on the estimate of F_{RA}.
- Safe flexural load-carrying capacity of the member (W_{bm}) corresponding to M_u

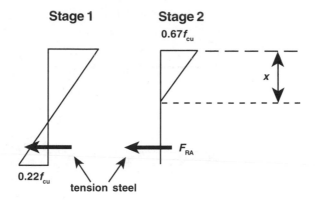

Figure 13.6 *Response of aircrete slabs to the applied load.*

and the span (2.5 m), and with an allowance for the own weight of the slab (1.44 kN).

The maximum anchorage force (F_{RA}) depends on the number of cross bars available between the section under consideration and the support (n_t), their diameter (ϕ_t), and the effective bearing stress (f_{ld}). A rule is adopted from the draft European Standard prEN12602 (1996), using cube strength in place of the cylinder strength and removing the partial factors for evaluating ultimate stage value of F_{RA}. The prEN rule evaluates the term F_{RA} separately for each tension bar and limits the effective bearing length (t_t) accordingly. It is proposed to modify this rule and evaluate F_{RA} and t_t corresponding to the entire tension steel.

$$F_{RA} = 0.83 n_t \phi_t t_t f_{ld} \quad \text{kN} \qquad \text{Eq. 13.1}$$

F_{RA} is also limited to '$0.8 n_t F_{wg}$', where F_{wg} is the declared strength of the welded connection (kN). The other terms are evaluated as follows:

$$f_{ld} = 1.8 m(e / \phi_t)^{1/3} f_{cu} \leq 2.7 f_{cu} \quad \text{N} / \text{mm}^2 \qquad \text{Eq. 13.2}$$

$$m = 1 + 0.3 \, (n_o/n_t)$$

where n_o denotes the number of bars in the compression zone close to the support (good detailing practice requires at least one bar at the support) and e is the distance from the central axis of cross-bars in the support zone from the nearest aircrete surface.

In all cases, critical value of f_{ld} was given as 10.26 N/mm² (2.7f_{cu}). Anchorage for tension steel was provided by this bearing pressure carried by four 6 mm dia. cross-bars between the applied load and the support (n_t =4), one bar being in the support zone (n_o = 1). Values of t_t are worked out as shown below, using the sketch given in Fig. 13.7. t_t is limited to 84 mm for a bar (14ϕ_t), where the average of spacings on either side of the bar exceeds 14ϕ_t .

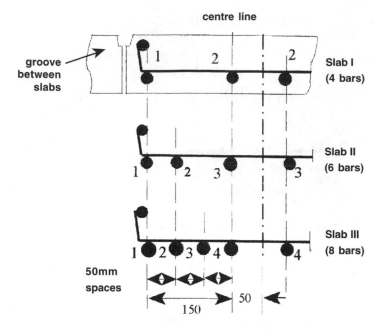

Figure 13.7 *Spacing of longitudinal bars (Leeds tests).*

slab type I (nos 1, 2 &3): $t_t = 4 \times 84$ = 336 mm
slab type II (nos. 4, 5 & 6): $t_t = 2 (42 + 25 + 75 + 84)$ = 452 mm
slab type III (no. 7, 8 & 9): $t_t = 2 (42 + 25 + 50 + 50 + 75)$ = 484 mm

Table 13.1 compares test loads at failure (W_f) with estimates of flexural load-carrying capacity (W_{bm}) of specimens. Factors of safety (FoS) are given as W_f / W_{bm}.

Estimated tensile stresses in steel in slab types I, II and III are 342, 306 and 246 N/mm^2, corresponding to the respective values of F_{RA}. These stresses are much lower than the yield strength 500 N/mm^2. Hence, the design should not be based on the yield strength of steel, as the provision of cross-bars may have to be limited to suit the detailing of reinforcement cage. Table 13.1 shows that the factors of safety for slabs are satisfactory, confirming the adequacy of the proposed design method. For design purposes, the load carrying capacities should be obtained by applying partial factors to the characteristic material properties, e.g., γ_c for aircrete as 1.50.

Shear capacity of aircrete slabs

The following rule is adopted from the European Standard prEN 12602 for the ultimate shear capacity of slabs (V_{Rd1}), with the partial factor for aircrete (γ_c) removed from the rule and with other modifications as given below:

$$V_{Rdl} = 0.0525[0.8f_{cu}]^{0.5}\left[\frac{1000-0.83d}{1000}\right]\left[1+240\frac{F_{RA}}{f_{st}bd}\right]bd/1000 \quad kN \quad \text{Eq. 13.3}$$

The prEN 12602 rule is modified to include an additional safety factor 1.2 (γ_n), by changing the first term from 0.063 to 0.0525. The original rule has a term $(1 + 240\rho_{eff})$, where ρ_{eff} is A_{eff}/bd and A_{eff} is the amount of fully anchored tension steel. Since effective anchorage is provided by the cross-bars, A_{eff} is taken as the area of steel that can sustain a force F_{RA}, with the stress reaching its yield strength f_{st} (500 N/mm²). This interpretation gives the term shown in Equation 13.3, $1 + F_{RA}/f_{st}bd$. Finally, the term $(1000 - 0.83d)/1000$ is limited to a maximum of 0.85, implying a limiting overall depth of about 200 mm for enhancement in shear provided by this factor.

Table 13.1 shows a comparison between loads at failure for slabs (W_f) tested at Leeds and the estimated values of shear capacity (W_s). W_s is given as $2V_{Rdl}$ less an allowance for the own weight of slab (1.44 kN). Table 13.1 shows that the proposed rule is satisfactory, as seen from the factors of safety W_f/W_s. This is also confirmed with the help of tests on 100 mm deep slabs (Table 13.2).

Anchorage of tension reinforcement

The information given in Table 13.1 shows the importance of cross-bars in providing anchorage to the longitudinal steel. This is apparent in the estimated loads and the test loads carried by slab types II and III, which are nearly the same despite a 33% increase in tension steel in slab type III over slab II.

Table 13.1 *Test results and estimates of flexural capacity (W_{bm}) and shear capacity (W_s), Leeds tests.*

Slab	F_{RA} (kN)	x (mm)	M_u (kNM)	W_{bm} (kN)	W_f (kN)	W_f/W_{bm}	W_s (kN)	W_f/W_s
1	68.67	90	10.03	22.63	28.4	1.26	20.22	1.40
2	68.67	90	10.03	22.63	24.7	1.09	20.22	1.22
3	68.67	90	10.03	22.63	24.4	1.08	20.22	1.21
4	92.38	121	12.53	28.64	34.1	1.19	22.00	1.55
5	92.38	121	12.53	28.64	34.0	1.19	22.00	1.55
6	92.38	121	12.53	28.64	33.0	1.15	22.00	1.50
7	98.92	130	13.14	30.09	32.0	1.06	22.49	1.42
8	98.92	130	13.14	30.09	33.4	1.11	22.49	1.49
9	98.92	130	13.14	30.09	33.3	1.11	22.49	1.48
Mean						1.14		1.42
Standard deviation						0.06		0.12

Effect of inadequacy of cross-steel was studied further with the help of tests carried out at the BRE, on 600 × 100 mm slabs. The load was applied through a loading frame, at two locations at quarter points on a span of 2.3 m (Fig. 13.8). The slabs had 6 nos. 8 mm dia. tension bars with clear cover of 20 mm. Compressive strength of aircrete was 4 N/mm², giving the corresponding value of modulus of elasticity as 150. The slabs were apparently obtained from cutting longer slabs. Unfortunately, this meant that extra anchorage bars in the support zone could be available only at one end of a test slab. The other end showed bare tension bars as cut from its longer parent slab and inadequate cross steel. Such slabs should not have been available for sale, as their use could lead to a potentially unsafe structure.

Tests on the slabs reported in this paper are divided in two categories. Slabs in the first category (slabs 1 and 2) were those obtained from cutting long slabs as described above. These slabs had only one 6 mm dia. cross-bar or U-shaped link effective between the support and the applied load. Slabs in the second category had two such bars (slabs 3, 4 and 5), with one in the support zone. In all cases, the critical value of f_{ld} was given as 10.80 N/mm² ($2.7f_{cu}$). The anchorage for tension steel (F_{RA}) was provided by this bearing pressure carried by the cross-bars between the applied load and the support. The longitudinal bars were evenly spaced and t_t was 480 mm for both types. The values of F_{RA} were 27.11 kN and 54.21 kN for the first and second type respectively.

The failure mode was similar to that for the tests at Leeds. The slabs failed close to the applied loads and, in case of slabs 1 and 2, on the side with deficient anchorage bars. Tensile stresses in steel corresponding to the respective values of F_{RA} for slabs in the two categories were 90 N/mm² (slabs 1 and 2) and 180 N/mm² (slabs 3, 4 and 5), much lower than the yield strength of steel and lower than those for the Leeds tests.

Table 13.2 shows estimates of neutral axis depth (x), moment of resistance (M_u) and flexural load-carrying capacity (W_{bm}) with an allowance of 1.6 kN to account for the weight of slab and the loading frame. Table 13.2 also shows test loads at failure (W_f) and factors of safety in respect of estimates of flexural and shear capacity of test slabs, similar to those in Table 13.1.

Serviceability design provisions (control on deflections)

Table 13.3 shows estimates of deflections in the test slabs (δ_p), for the three types of specimens tested at Leeds marked as I, II and III. The estimated loads (W_{def}) correspond to the limit of the elastic stage response of the slabs. The tensile stress at the soffit is limited to $0.22f_{cu}/\gamma_c$, with γ_c taken as unity for deflection purposes. Calculations are based on the moment of inertia of uncracked section (MoI) and the equivalent area of tension steel. Table 13.3 compares estimated deflections with the values (δ_m) read from load-deflection graphs against the loads W_{def}. The load-deflection relationship was linear and it continued marginally beyond the estimated loads W_{def} for all slabs, showing that the slabs behaved elastically for this level of loading.

The BRE tested two additional specimens, similar to slabs 3 to 5 of the BRE tests

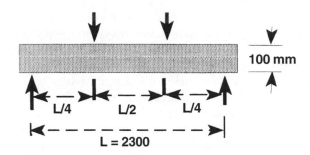

Figure 13.8 *Schematic loading (BRE tests).*

reported earlier. For the first slab, the applied load represented a design load of 2.1 kN/m²
and the second slab was tested with a load of 3.5 kN/m² to study implications of an extra
load due to ponding on a roof. The deflections for the two types were 4.22 mm and 6 mm
under the applied total loads of 4.6 kN and 6.4 kN respectively. These values increased
by 25% for the first slab and 31% for the second slab, confirming that the creep effect
increases with higher stresses in aircrete.

Table 13.2 *Test results and estimates of flexural capacity (W_{bm}) and shear
capacity (W_s), BRE tests.*

Slab	F_{RA} (kN)	x (mm)	M_u (kNM)	W_{bm} (kN)	W_f (kN)	W_f / W_{bm}	W_s (kN)	W_f / W_s
1	27.11	34	1.76	6.11	8.4	1.38	8.02	1.05
2	27.11	34	1.76	6.11	9.3	1.52	8.02	1.16
3	54.22	67	2.90	10.09	11.6	1.15	10.16	1.14
4	54.22	67	2.90	10.09	11.6	1.15	10.16	1.14
5	54.22	67	2.90	10.09	12.4	1.23	10.16	1.22
Mean						1.29		1.14
Standard deviation						0.14		0.06

Table 13.3 *Estimates of deflections of 600 × 200 mm slabs (Leeds tests).*

Slab type	A_{st} mm²	x (mm)	MoI (mm⁶)	W_{def} (kN)	δ_p	δ_m
I	201	116	545	11.8	4.2	3.8
II	302	122	597	13.8	4.5	4.0
III	402	126	640	16.0	4.8	3.8

The BSI Draft prEN 12602 (1996) allows checking deflections under the quasi-permanent load condition, which gives the total load as $(G_k + \psi_2 Q_k)$. G_k is the permanent load and ψ_2 is 0.3, the coefficient for combination value for a quasi-permanent variable action Q_k or, in this case, the imposed load. For normal design situations, this load condition would give bending moments conveniently within the elastic range for aircrete slabs, so that the limiting tensile stress at the extreme fibre of the composite section is $0.22 f_{cu}/\gamma_c$. ($\gamma_c = 1$, for checking deflection)

Deflections should be limited to span/250 under the quasi-permanent load condition, in the interest of appearance and general utility of the floors. In roof slabs, the limit should be span/360 and the spans should be limited to 4.0 metres. Slabs should be laid with a slope of 5%, spanning at right angles to the slope to facilitate the drainage. Deflections may increase and reach critical level of span/100, because of creep and weakening of aircrete, if it has come in contact with moisture. In such cases, roof planks should be replaced and the protective membrane should be reinstated.

Choice of the ultimate stage design and serviceability design

Floor slabs are invariably provided with a screed to protect the aircrete surface from abrasion. The screed is often reinforced and bonded to the aircrete slabs to provide robustness to the floor and to achieve continuity at supports. The tension steel at support should have satisfactory bond with the dense screed. The moment of resistance at support sections, therefore, should depend on critical compressive stress in aircrete. At midspan, the depth of the screed should be adequate in most cases to provide a compression flange. The design should be based on balancing the compression with tension depending on the anchorage provided by cross-bars to the tension steel, as described earlier. For most design situations, it should be possible to limit the deflection to be less than span/250 with the depth of the slab required for strength purposes and with the relief provided by continuity of the slab over the supports. Further research is required in the field of effectiveness of a bonded screed in providing composite action with aircrete, for the purposes of strength and serviceability methods of design.

Roofs are subjected to imposed loads, which are normally much less than the floor loads. Depth of an aircrete slab, therefore, should be adequate to satisfy the limit on deflection for roofs (span/360). However, roof slabs are often simply supported over rafters, protected from the elements with a membrane and not with a screed. Under these circumstances, designers may wish to consider limiting the design to the elastic stage, with partial factors allowed by the Eurocode as appropriate to the quality control measures in manufacturing process of precast products. This procedure will provide a streamlined and coherent design, in harmony with the actual behaviour of the material.

Figure 13.9 shows limiting depths of roof slabs with 0.25% tension steel and spans up to 4.0 metres. The applied load is the characteristic imposed load per square metre. The design is based on limiting deflection to span/360 for an applied load of $G_k + 0.3Q_k$.

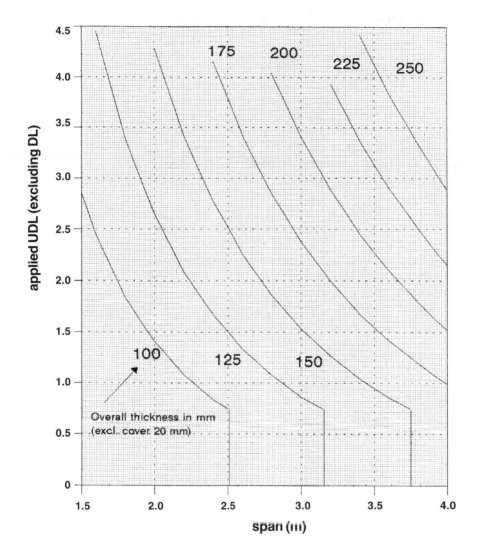

Figure 13.9 *Load-span charts for aircrete roof slabs with 0.25% steel.*

Tensile stress is limited to $0.22f_{cu}/1.5$, to resist the bending moments produced by the applied load of $1.35G_k + 1.5Q_k$, adopting the normal value of γ_c as 1.5 and ignoring any reduction allowable with quality control measures, etc. Manufacturers may wish to use a reduced value of γ_c and prepare graphs with other percentages of steel.

Conclusions

The proposed method is intended to have the design consistent with the nature of aircrete, particularly its very low compressive strength and a higher ratio of tensile strength to compressive strength in comparison with concrete. It could also avoid detailed calculations for curvature, cracked properties, etc. for deflection purposes. For estimating load carrying capacities, further research is required to achieve a reliable bond between the protective coating and the steel. Analytical and experimental work is also essential for improved anchorage details to achieve effectiveness of the reinforcement. Further studies are required in the field of effectiveness of a bonded screed in providing composite action with aircrete, for the purposes of strength and serviceability methods of design.

Further research should be carried out to investigate feasibility of using nonferrous reinforcement in aircrete slabs, e.g., carbon-fibre rods. In some cases, strength and modulus of elasticity of such products may be less than the corresponding values for steel. However, tensile stresses in reinforcement in aircrete slabs would be low as shown by the estimates given by the proposed design method. Properties of carbon-fibre could, therefore, be quite acceptable. The main advantage will be in the serviceability of the roofs and floors, with elimination of problems associated with rusting of steel. This concept could become increasingly attractive with further developments in the manufacturing process of carbon fibres, leading to its lower costs and increased use in various other practical applications.

References

British Standards Institution, 1985. BS8110: *Structural Use of Concrete*, Part 2: Code of practice for special circumstances.

British Standards Institution, 1996. Draft prEN 12602 : *Prefabricated reinforced components of autoclaved aerated concrete*.

British Standards Institution, 1997. BS8110: *Structural Use of Concrete*, Part 1: Code of practice for design and construction.

Building Research Establishment, 1996. *IP 10/96 Reinforced Autoclaved Aerated Concrete Planks Designed before 1980*, CI/SfB (2) H q6.

RILEM 1993. Technical Committees 78-MCA and 51-ALC, *Autoclaved Aerated Concrete – Properties, Testing and Design.* London:E & FN Spon.

Acknowledgements
The author would like to thank the Association of Autoclaved Aerated Concrete Association for their sponsorship of the tests. These tests were carried out under the guidance of Professor A W Beeby and Professor R Narayanan of the University of Leeds and their help is gratefully acknowledged. The author would also thank the Building Research Establishment for the information on testing and inspection of aircrete slabs and the Department of the Environment, Transport and the Regions.

14 Lessons learned from the Oklahoma City bombing

W. Gene Corley

The truck bombing of the Murrah Building on April 29, 1995 caused significant damage to the structure. From the characteristics of the bomb crater, it was determined that the explosion yielded energy comparable to that from the detonation of 1,814 kg (4,000 lbs.) of trinitrotoluene (TNT).

The following describes collection of data, engineering details of the building, and the failure mechanism. Calculations show the mechanism caused by removal of one or more columns along the north side of the building would lead to extensive progressive collapse. Possible details for improved performance are investigated.

The blast directly removed a principal exterior column and the associated airblast caused failure of two others. The airblast also destroyed some of the floor slabs in the immediate vicinity. From visual inspection and analysis of the damage, it is shown that collapse extended the damage beyond that caused directly by the blast. Damage that occurred and the resulting collapse of nearly half the building is consistent with what would be expected for an ordinary moment frame building of the type and detailing available in the mid-1970s, when subjected to the blast from the large truck bomb.

Using information developed for North American building codes, types of structural systems that would provide significant increases in toughness for structures subjected to catastrophic loading from events such as major earthquakes and blasts are identified. One of these systems is compartmentalized construction in which a large percentage of the building has structural walls that are reinforced to provide structural integrity should the building be damaged. Two additional types of detailing, used in areas of high seismicity, are special moment frame construction and dual systems with special moment frames (herein referred to as dual systems).

These systems are shown to provide the mass and toughness necessary to reduce the effects of extreme overloads on buildings. Consequently, it is recommended that these structural systems be considered where a significant risk of seismic and/or blast damage exists.

Introduction

Corley *et al.* (1996, 1998) describe the nine-storey office building of the Murrah Building project, its structural system, and other aspects of the project.

Failure mechanisms for the building and engineering strategies for reducing damage from blast to new and existing buildings are explored. Specifically, mechanisms for multi-hazard mitigation, including wind and earthquake effects, are considered. Among the strategies evaluated are procedures and details from the Federal Emergency Management Agency (1995).

The Murrah Building project was designed for the Design & Construction Division, Region 7, Fort Worth, Texas, of the GSA Public Building Service, Washington, DC (BPAT, 1996; Corley *et al.*, 1996, 1998). Figure 14.1 shows the site plan along with the column lines. The project consisted of the nine-story Murrah office building (hereafter referred to as the nine-storey portion of the Murrah Building) with one-storey ancillary east and west wings. An adjacent multi-level parking structure, partially below grade and partially above grade, was located south of the office building.

Figure 14.1 shows the entire project located between North Harvey Avenue on the West, N. W. Fifth Street on the north, North Robinson Avenue on the east, and N.W. Fourth Street on the south. This chapter focuses on the nine-storey portion of the Murrah Building, which incurred significant damage and partial collapse as a result of the April 19, 1995 bombing, The nine-storey and one-storey portions of the building were demolished after search and rescue were completed. The parking structure was barely damaged.

The Building Performance Assessment Team provides a full description of the design, construction, and condition of the Murrah Building prior to the tragic bombing (BPAT, 1996). The structure was a reinforced concrete frame with three rows of columns spaced at 6.1 m (20 ft) within each row. A large transfer girder at the third floor permitted the elimination of alternate exterior columns below. The building was designed and constructed in accordance with the applicable codes, but did not provide any deliberate resistance against a vehicular bomb attack. Additional information is given in companion papers (Corley *et al.*, 1998; Sozen *et al.*, 1998).

An estimate is made of the blast loading and its direct effect on the structure of the building. First, the blast loading is calculated from the properties of the crater formed by the explosion. Then, the response of critical structural elements to the calculated loading is determined using approximate methods appropriate for this assessment. Critical elements include the principal exterior columns supporting the transfer girder and the floor slabs of the building.

Of particular interest is the integrity of the structural frame with these damaged columns. Strategies to mitigate similar damage in other buildings are discussed elsewhere (Corley *et al.*, 1996).

Scope of work

Investigation of the area around the Murrah Building in Oklahoma City took place during the period of May 9 through 13, 1995, 3 weeks after the blast occurred on Wednesday, April 19. The location of the Murrah Building in downtown Oklahoma City is shown in Fig. 14.2.

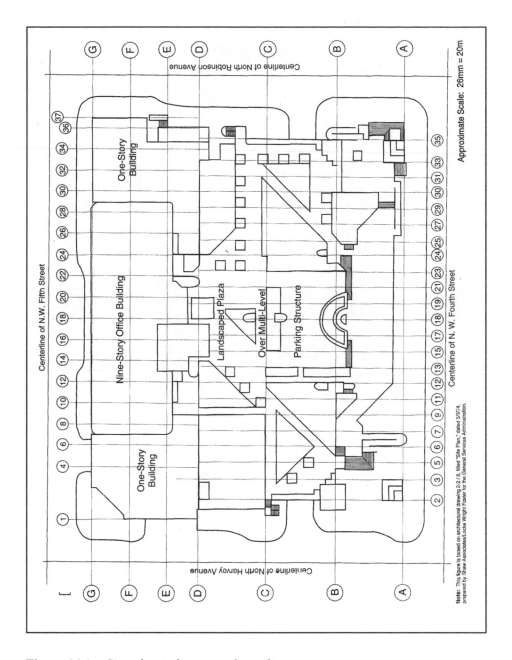

Figure 14.1 *Site plan indicating column lines.*

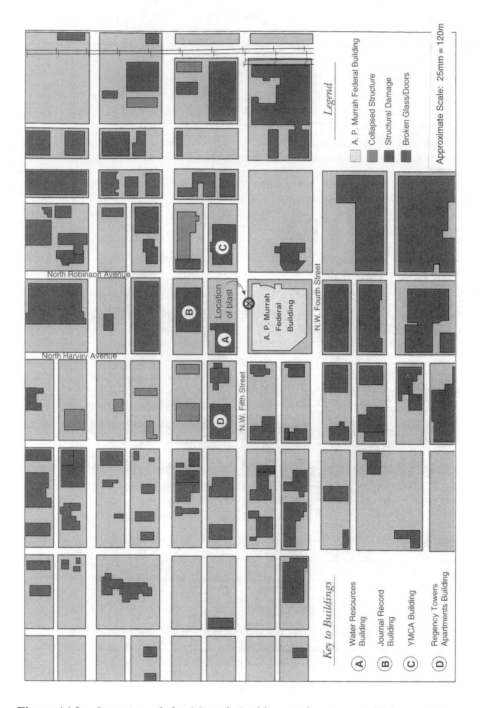

Figure 14.2 *Location of the Murrah Building in downtown Oklahoma City.*

While in Oklahoma City, the investigators took photographs; collected structural drawings, shop drawings, photographs, and samples of structural components, including concrete and reinforcing bars; and obtained an audio tape of the blast. The team also conducted interviews to obtain information concerning damage to buildings. Physical inspection of the structure was limited to visual observation from a distance of approximately 200 ft.

Following the site visit, samples of concrete and reinforcing bars taken from the site were tested to determine physical properties of materials used in the building. Work performed included developing the most probable response of the building to the blast and determining whether new technology can be used to enhance the resistance of buildings to blast, wind, earthquake, and other hazards.

Data collection

Concrete and reinforcing bar samples

On May 12, 1995, the BPAT visited the Oklahoma County Sheriff's Firing Range to interview personnel and view debris from the Murrah Building. During this visit, photographs taken soon after the explosion by several law enforcement organizations were reviewed. Also, several pieces of building debris were inspected.

Inspection of debris disclosed that there were a few 'chunk' samples, sections of spandrel beams, some large slab sections, and a few pieces of deformed reinforcing bar and it was confirmed that the concrete and the reinforcing bars had come from the Murrah Building. During the visit, locations were marked on six concrete debris samples where cores were to be taken. In addition, a seventh sample, a chunk of concrete, was also marked for coring. Several sections of reinforcing bars were marked to be taken as samples.

On May 12, 1995, a concrete coring company took six 152-mm (6-in) cores from the marked areas. After the cores had been taken in the field, they were packed in a plastic cooler and shipped to Construction Technology Laboratories, Inc. (CTL), in Skokie, Illinois where selected samples were tested. Similarly, reinforcing bar samples were put in a heavy plastic shipping tube and sent to CTL for testing. All samples arrived at CTL on May 15, 1995. Chain-of-custody documentation was maintained for all samples.

Photographs and video tapes

In addition to conducting a visual inspection, the team reviewed photographs and video tapes recorded on April 19, 1995, and during the following rescue and recovery period. These records were useful in establishing the performance of the building.

Audio tape

At the time of the blast, a hearing on water rights was convened in the Water Resources building, located at the northeast corner of North Harvey Avenue and N.W. Fifth

Street, across the street and about half a block from the centre of the explosion (*see* Fig. 14.1). An audio tape recorder, used to document the hearing, was started approximately two minutes before the blast occurred.

The sound of the blast was captured on the audio recorder. As can be seen from the analysis of the recording (Fig. 14.3), the blast was followed by a brief period of moderate noise, then a period of about three seconds during which the noise level exceeds the range of the recorder. The 3-second period is interpreted to be the time it took for the building to collapse. This observation is consistent with findings concerning the collapse mechanism.

Interviews

In a comprehensive effort to obtain information, members of the team interviewed several key individuals and groups of people. Interviews were conducted in an effort to determine the original design parameters for the building and to obtain information about events as the building responded to the blast loading. Also information was obtained about the damage present in the building immediately after the collapse.

Structural engineer of record

An interview with the structural engineer of record for the Murrah Building project was done on May 11, 1995. He noted that the original design parameters did not require any

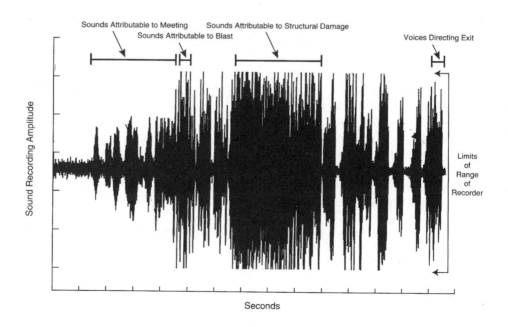

Figure 14.3   *Analysis of the recording of the sound of the blast.*

consideration of resistance to blast, earthquake, tornado, or other extreme loading. Rather, the structure was designed for normal office building loading in Oklahoma City.

It was stated that soon after the blast, the structural engineer was called to the site to assist with stabilizing the debris. He noted that the primary failure involved collapse of the north portion of the office structure. Two bays in the south half of the building also collapsed. No damage was observed that suggested significant lateral movement of the structure. He saw no visible signs of hinging at the tops or bottoms of any remaining columns. Also, there were no visible indications that shearwalls had hinged.

In general, the structural engineer of record believed that the collapse was initiated by failure of columns along the north side of the building. In his opinion, two or more columns may have been destroyed in the vicinity of where the blast originated. He felt that perhaps a third column supporting the transfer girder and one interior column could have been destroyed by the blast itself.

Water Resources Board employee

An employee of the Oklahoma State Water Resources Board, who was in the Water Resources building at the time of the blast, was interviewed on May 11, 1995. He stated that when the explosion occurred, he noticed the building shaking, fluorescent lights and windows exploding, ceiling tile falling, and walls blowing down. Immediately after the blast, he smelled gas and nitrates.

Although the water resources employee was knocked to the floor and was aware that there was substantial damage to his surroundings, he did not 'hear' the explosion. After helping other employees leave, he left the Water Resources building and walked about one block northwest to where he had parked his truck. Even though it was parked about two blocks from the explosion, the truck had sustained severe damage including dented doors and broken windows.

Structural engineer

A structural engineer from Oklahoma City, who was on the site immediately after the blast, provided very helpful insight concerning damage to surrounding structures. He was called in to inspect a number of buildings damaged by the explosion. These included the 24-storey Regency Towers apartment building that has a reinforced concrete frame; the Journal Record building that also has a reinforced concrete frame; the one-storey Journal Record Annex, which has a light-metal steel frame; and the YMCA building that has a reinforced concrete frame and is across the street from and northeast of the Murrah Building site. In addition, a number of low-rise, unreinforced masonry buildings and steel-frame buildings in the area were inspected.

The local structural engineer noted that although there was non-structural damage to many buildings, he found little significant structural damage to engineered structures more than two blocks away from the Murrah Building. Specifically, the high-rise Regency Towers apartment building, the main multistorey portion of the Journal Record building, and the YMCA building had no damage to their structural frames. Several non-load-bearing walls inside those three buildings and portions of the roofs were damaged or destroyed, but the structural frames were intact and essentially undamaged. None of these buildings was in danger of general collapse.

It was noted that the one-storey light-metal steel building referred to as the Journal Record Annex was badly damaged. There was severe damage to exterior walls and the roof, and the loading dock enclosure was near collapse. Also, wood frame, bowstring truss and light metal buildings within two blocks were damaged or had collapsed.

GSA structural consultant

A Texas firm that serves as structural consultant to GSA, the owners of the building, provided information to the investigators.

A representative of the Texas firm noted that failure surfaces he observed did not indicate that shear strength controlled the failure of the slabs and beams. Most surfaces were at a slope that is generally associated with flexural behaviour. He noted that he had not been able to observe the failure patterns for destroyed columns G-16, G-20, and G-24 in the lower two stories.

When asked about the conditions of columns still standing, the GSA structural consultant stated that there was no indication the columns had hinged at the tops and bottoms, as would be the case if the building developed an overall hinging mechanism because of lateral load. Rather, he noted that the general mechanisms that developed were associated with loss of one or more columns due to blast loading.

FEMA engineering consultant

An engineering consultant working for the Federal Emergency Management Agency provided the team with his observations. This consultant also confirmed there was no evidence of flexural hinging at the tops and bottoms of columns. Consequently, he concluded the building did not develop a failure mechanism from lateral load.

Based on his observations, the FEMA consultant offered the opinion that at least one column near the blast had been destroyed by brisance (the shattering effect of the blast). He also believed that at least one and perhaps two other columns had been knocked out by the blast.

Tests of materials

Concrete

Five concrete cores taken from the Murrah Building were selected for compression testing. In addition, two cores were taken from the chunk sample; one of these was tested in compression and the other was subjected to petrographic examination (Corley, 1996).

Although there was evidence of damage from blast and/or from handling stresses, visual inspection of the concrete indicated that it met or exceeded the quality called for in the design specifications. Compression test results indicated that the concrete strength was well in excess of the 28 MPa (4000 psi) called for in the design specifications.

Petrographic evaluation indicated that the concrete contained normal-weight aggregate and was of the quality required in the design specifications (ibid.).

Reinforcing bars

Several pieces of reinforcing bars were recovered from the debris of the building. A few lengths of straight bar were tested in tension (for results, see Corley *et al.*, 1996). Tests showed the yield stresses and strengths of the bars were greater than the minimums specified and exceeded the requirements of the design specifications.

Structural drawings and system

Structural drawings show that the Murrah Building consisted of cast-in-place ordinary reinforced concrete framing with conventionally reinforced columns, girders, beams, slab bands, and one-way slab system. Exterior spandrels supporting the exterior curtain wall were exposed concrete with a vertical-board-formed finish. The primary lateral load resisting system for wind forces was composed of reinforced concrete shear walls located within the stair and elevator system on the south side of the building. Although neither the governing building code nor the owner required consideration of blast loading, earthquake loading, or tornado in the design, the required wind-load resistance provided substantial resistance to lateral load.

According to general notes on the structural drawings, the Murrah Building project was all reinforced concrete. It was proportioned, detailed, fabricated, and delivered in accordance with the American Concrete Institute Building Code Requirements for Reinforced Concrete (ACI 1971). The specified yield strength for ties, #3 bars, and stirrups was 276 MPa (40,000 psi). The specified yield strength for all other deformed bars and all welded wire fabric was 414 MPa (60,000 psi). Specified 28-day concrete compressive strength was 21 MPa (3000 psi) for foundation and equipment bases and 28 MPa (4000 psi) for the structural beams, slabs, columns, walls, pilasters, spread footings, and parking garage exterior walls. As indicated in the general notes, all reinforcing bar splices were to be lapped 30 bar diameters unless otherwise noted. Design live loads for the project are listed in Table 14.1.

The blast occurred on the north face of the nine-storey portion of the Murrah Building. This account describes the configuration of the structural system on the north face, along column line G, as well as the intermediate columns along the column line F, primarily in the area between column lines 16 and 28.

All 117 architectural drawings and 40 structural drawings for the project are carefully detailed, well prepared, and well coordinated. The level of structural detailing (detailing is the process of selecting and designating on drawings the amounts, lengths, bends, and locations of steel reinforcement in reinforced concrete) and the use of schedules with full dimensions for all slabs, T-beams, spandrel beams, transfer girder, and column reinforcing bars are significantly better than normally expected for buildings of this type. Although a complete structural design check was not undertaken, those components that were reviewed were found to comply with accepted standards and ACI 318-

Table 14.1 *Design live loads for project.*

Project element (1)	Load (Pa) (2)
Roof	960
Office areas (excluding 20 psf load provision for movable partition)	2400
Parking structure floor	2400
Corridors, stairs and lobbies	4800
Plaza pedestrian walks	4800
Mechanical equipment spaces, elevator machine rooms, equipment load	7200
Vehicle maintenance	7200
Street-level sidewalk and approaches	12000

Note: blast and earthquake loads were not prescribed by the building code and were therefore not considered.

71 (1971). A spot check of the reinforcing bar shop drawings showed compliance and good correlation with the structural contract documents.

Selected structural details and dead load analysis

The following discussion is based on a review of the structural drawings for the Murrah Building. Figures 14.4 to 14.6 show the structural layout.

Typical floor framing

One-way concrete slabs (identified as TS on the structural drawings) spanned the east-west direction. These slabs were 153-mm (6-in) thick and were reinforced with #4 bars 5330 mm (17 ft 6 in) long at 457mm (18 in) on centre and #4 bars 3660 mm (12 ft) long at 457 mm (18 in) on centre. The amount of reinforcing at the centreline of the span was #4 bars at 229 mm (9 in) bottom steel or 174 sq mm (0.27 sq in) of 414 MPa (60,000 psi) yield reinforcing bar per 300 mm (12 in) of width. The longer bottom bars extended 229 mm (9 in) into the supporting T- beams at each end.

Top slab reinforcement centred over the beams on the column lines consisted of #4 bars 3048 mm (10 ft) long at 406 mm (16 in) on centre and #4 bars 3658 mm (12 ft) long at 406 mm (16 in) on centre. Therefore the amount of reinforcing was #4 bars at 203 mm (8 in) on centre at the centreline of the support, or 194 sq mm (0.30 sq in) of reinforcing bar per 300 mm (12 in) of width.

The one-way slabs spanned 6096 mm (20 ft) centre to centre of supports in the east-west direction and were supported by deeper slab bands (T-beams) that were 1219 mm (48 in) wide by 510 mm (20 in) deep. Therefore, the face-of-support to face-of-support span for the one-way slabs was 4880 mm (16 ft). The longer of the top bars in the slab

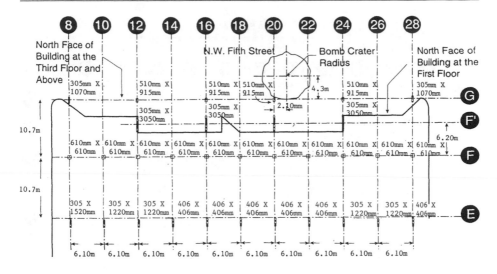

Figure 14.4 *Structural layout – column locations and dimensions (first floor).*

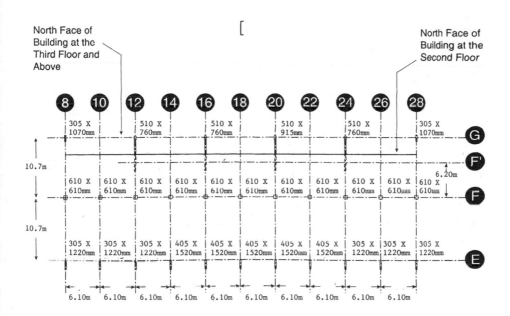

Figure 14.5 *Structural layout – column locations and dimensions (second floor).*

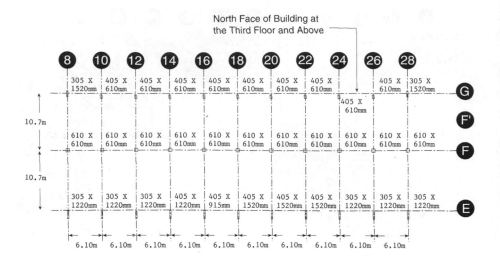

Figure 14.6 *Structural layout–column locations and dimensions (third floor through roof).*

extended 1220 mm (4 ft) into the slab span outside of the T-beam. As a result, there was no top steel in the slabs for the middle portion of the 4880 mm (16 ft) slab spans.

Slabs contained minimum steel in the north-south direction consisting of #4 bars at 457 mm (18 in) on centre (290.32 mm sq/m or 0.135 sq in/ft). This steel was placed directly on top of the bottom main east-west reinforcing bars. One of the structural drawings contains the note referring to minimum steel. Shop drawings that show how far the minimum steel extended into the spandrels beam at the north face of the building or how much lap occurred at splices were not available.

The T-beams at the third floor and the typical floor framing in the north-south direction spanned 10,668 mm (35 ft) and as reported above, measured 1220 mm (48 in) wide by 510 mm (20 in) deep. Reinforcing in these beams is listed in Table 14.2.

There was no continuous top steel along the span of the T-beams from column line G to column line F. The top of the middle 3810 mm (12 ft 6 in) portion of the span was unreinforced. Lack of continuous reinforcement was consistent with detailing required for ordinary moment frame buildings. However, it is noted that a plane of weakness existed where main reinforcement was terminated.

At each typical floor (4 to 9) there was an east-west spandrel beam along the north face that measured 457 mm (18 in) wide by 889 mm (35 in) deep. Reinforcing in these beams is listed in Table 14.3 and shown in Fig. 14.7.

Spandrel beams had at least two continuous # 8 reinforcing bars at the top with a lap splice of 4267 mm (14 ft), resulting in total cross-sectional area of tensile reinforcement for negative moment of two # 8 bars (1019 sq. mm or 1 .58 sq. in) plus two # 9 bars (1290 sq. mm or 2.0 sq. in), or a total of 2342 sq. mm (3.58 sq. in).

Table 14.2 *Reinforcement of third-floor beams.*

Column lines	T-beam	Location	Number/size	Length (m)
10,12, 24 and 26	6	Bottom (short)	2-#7	7
		Bottom at column (long)	4-#8	10
		Top at column line G	4-#10	3.5
		Top at column line F	4-#11	3.75
		Stirrups	#4 at 225mm	
16, 18 and 20	12	Bottom (short)	2-#7	6.75
		Bottom at column (long)	4-#8	10
		Top at column line G	3-#11	3.25
		Top at column line F	3-#10	7
		Stirrups	#4 at 225mm	
14 and 22	10	Bottom (short)	2-#7	6.75
		Bottom at column (long)	4-#8	10
		Top at column line G	2-#10	3.25
		Top at column line F	4-#11	7

Table 14.3 *Reinforcement of spandrel beam along north face.*

T-beam	Location	Number/size	Length (m)
18	Bottom at centreline	2-#7	9
	Bottom at column	2-#8	12.75
	Side face	2-#5	6.75
	Top at line	2-#8	10.25
	Stirrups	#6 at 330mm	
	Stirrups	#4 at 330mm	
19	Bottom at centreline	see T-beam 18	
	Side face	2-#5	6.75
	Top at column line	2-#9	7.25
	Stirrups	#6 at 330mm	
	Stirrups	#4 at 330mm	
20	Bottom at centreline	2-#7x	9
	Bottom at centreline	2-#8x	12.75
	Side face	2-#5x	6.75
	Top at column line	2-#8x	10.25
	Stirrups	#6 at 330mm	
	Stirrups	#4 at 330mm	

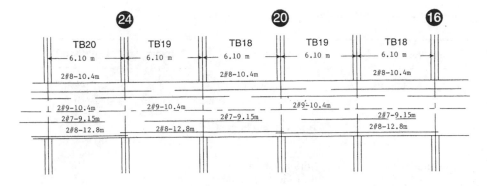

Figure 14.7 *Reinforcing in east-west spandrel beam along the north face.*

Column sizes along column line G from the third floor to the roof were typically 406 mm by 610 mm (16 in by 24 in). Reinforcement consisted of four # 9 vertical bars at the lower levels and four # 8 vertical bars at the upper levels, with #3 horizontal ties at 406 mm (16 in) on centre.

Third floor framing

In order to enhance the street level appearance and open the building to allow access, the designers introduced transfer girders along column line G at the third floor. These girders spanned 12,192 mm (40 ft), thus supporting the columns located at the column lines 10,14,18,22, and 26. Columns below were spaced 12,192mm (40 ft) on centre and were therefore located at column lines 12, 16, 20, and 24. These columns measured 508 mm by 914 mm (20 in by 36 in) with the 914 mm (36 in) dimension in the north-south direction.

Reinforcement for these columns consisted of 20 #11 vertical bars and # 4 horizontal ties at 254 mm (10 in) on centre. The transfer girders at the third floor (with designations 3B-3 and 3B-4 on the structural drawings) measured 914mm (36 in) wide by 1524 mm (60 in) deep and contained heavy reinforcement (Fig. 14.8). Reinforcing in these girders is listed in Table 14.4.

Bottom steel in the transfer girders was configured such that the longest bars (4 #11) were 12,192 mm (40 ft) long. Thus no lap occurred at the support columns at column lines G12, G16, G20, and G24, a detail typical at the time this structure was built. Top steel in these transfer girders was such that the 3 #11 continuous top bars left a 1524 mm (5 ft) lap at the centre of the 12,192 mm (40 ft) span at column lines 14,18, 22, and 26.

Second floor framing

The edge of the slab of the second floor was set in approximately 2996 mm (9 ft 6 in) from column line G. Spandrels beams 2B-33 and 2B-34 had cross sections 457 mm by 1219 mm (18 in by 48 in) and spanned12,192 mm (40 ft). These spandrels picked up

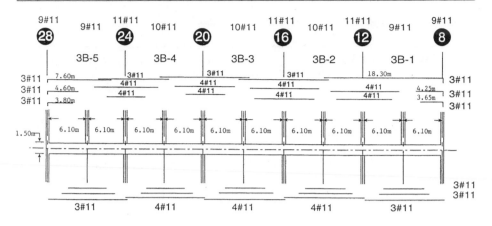

Figure 14.8 *Transfer girders at the third floor.*

Table 14.4 *Reinforcement of third-floor transfer girders.*

Column line	Girder	Location	Number/size	Length (m)
between 16 and 20	3B-3	Bottom at centreline	3-#11	6
			3-#11	8.5
			4-#11	12.25
		Top at column line	4-#11	6
			4-#11	9.25
			3-#11	13.75
		Stirrups	#6 at 152mm	
between 20 and 24	3B-4	Bottom at centreline	same as 3B-3	
		Top at column line	4-#11	6
			4-#11	10
			3-#11	13.75
		Stirrups	#6 at 152mm	

portions of the facade load, the third floor slab load, and the north end reaction of the second floor beams 2B-22 and 2B-17. Table 14.5 lists reinforcement. Top steel at the centreline of the span consisted of 2#6 bars with a lap splice of 762 mm (2½ft) and 3 #9 bars. These spandrels beams were supported by a concrete wall column 305 mm (12 in) wide by about 3048 mm (10 ft) long with 24 #7 vertical bars and #4 horizontal bars at 305 mm (12 in) in the centre.

Table 14.5 *Reinforcement of second-floor spandrels.*

Beam	Location	Number/size	Length (m)
2B-33	Bottom at span centreline	2-#7	8
		2-#8	12.5
	Top at column line	3-#9	7
	Top at span centreline	2-#6	7
	Sttirrups	#5 at 560mm	
2B-34	same as 2B-33		

Table 14.6 *Reinforcement of roof beams.*

Beam	Location	Number/size	Length (m)
RB-4	Bottom short	2-#8	6
	Bottom at column (long)	3-#8	10.25
	Top at column line G	4-#8	3

Roof structure

Roof slabs (designated RS-2 on the structural drawings) spanned in an east-west direction and were 152 mm (6 in) thick, 4877 mm (16 ft) clear span, and reinforced with #4 bottom bars at 406 mm (16 in) on centre. Total cross-sectional area of tensile reinforcement at the centreline was 323 sq. mm/m (0.15 sq. in/ft) of slab width. The bottom bars extended 229 mm (9 in) into the supporting beams (designated RB at each end on the structural drawings).

Slab top steel centred over the column lines consisted of #4 bar 3048 mm (10 ft) long at 610 mm (24 in) on centre and #4 bars 3810 mm (12½ft) long at 610 mm (24 in) on centre. Therefore, the amount of top reinforcing was #4 bars at 305 mm (12 in) on centre at the centreline of the support, or 430 sq. mm/m (0.20 sq. in/ft) of slab.

The roof reinforcing configuration and dimensions are the same as for the typical floors. The roof beams (designated RB on the structural drawings) spanned 10,668 mm (35 ft) in the north-south direction and were 1219 mm (48 in) wide by 508 mm (20 in) deep. Reinforcing in these beams is listed in Table 14.6.

No continuous top steel was provided across the span of the roof beam from column line G to column line F. The middle 3810 mm (12½ ft) portion of the span was unreinforced at top of the roof beams. Spandrel beams in the east-west direction were 457 mm (18 in) wide by 1194 mm (47 in) deep and contained the reinforcing listed in Table 14.6 (*see* Figs. 14.9 and 14.10).

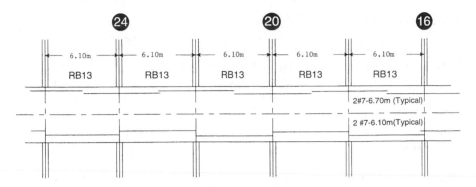

Figure 14.9 *Roof beams in east-west direction.*

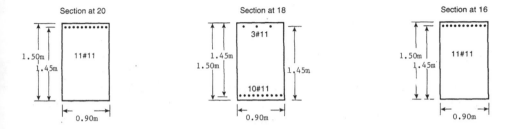

Figure 14.10 *Spandrel beams at third floor in east-west direction.*

Roof spandrel beam reinforcing was different from the typical floor. There was no lap at the column line in the bottom steel, and there was continuous top reinforcing with a 610 mm (2 ft) lap at the centreline of the span. Lack of a lap in the bottom steel creates a plane of weakness when unanticipated load such as wind, earthquake, and blast induced forces cause positive moment at columns.

Table 14.7 *Reinforcement of interior column.*

Floor	Vertical bars (number/size)	Ties (number and spacing)
3-4	16-#11	#4 at 400m on centre
4-5	12-#11	#4 at 400m on centre
5-6	8-#10	#3 at 400m on centre
6-7	4-#10	#3 at 400m on centre
7-8	4-#9	#3 at 400m on centre
8-9	4-#9	#3 at 400m on centre
9-roof	4-#8	#3 at 400m on centre

Interior columns

Interior column F22 was 610 mm by 610 mm (24 in by 24 in) from the ground up to the third floor. Reinforcement consisted of 16 #11 vertical bars at the first floor, and 12 #11 vertical bars at the second floor with #4 horizontal ties at 406 mm (16 in) on centre. From the third floor to the roof, the column was 508 mm by 508 mm (20 in by 20 in) and contained reinforcing shown in Table 14.7.

Foundation

The building foundation is a drilled pier system with allowable soil bearing pressure for dead load plus 50 percent live load of 1.4 MPa (30,000 psi) at elevation 1194 to 1200 m.s.l., and 2.4 MPa (50,000 psi) at elevation 1175 to 1194 m.s.l. Lengths of drilled piers shown on drawings range from 6096–9144 mm (20–30 ft).

Drilled piers supporting the building have large belled bottoms with diameters vaiying from 610 mm to 2438 mm (24 in to 96 in). Smaller shaft diameters vary from 457 mm to 1219 mm (18 in to 48 in).

Dowels

All columns are dowelled into the top of the pier below. At the 508 mm by 9144 mm (20 in by 36 in) columns, eight #10 dowels extend into a 1219 mm (48 in) diameter drilled pier, and at the 305 mm by 3048 mm (12 in by 120 in) wall column, four #7 dowels extend into each 610 mm (24 in) diameter drilled pier, one at each end.

Lateral load analysis

Original design

Wind loads were carried primarily by the core/shear wall assemblage at the middle of the south face of the building. This system conformed to the code and was appropriate for a nine-storey building with wind pressures of 1197–2154 kPa (25–45 psf).

The total wind load in the east-west direction is approximately 149,685.48 kg (330 kips, where 1 kip = 1000 lb). Had the original design assumed that one half of the east-west wind load was resisted by the frame in the north face, then 74,800 kg (165 kips) would have been the portion of the wind load. Simple hand calculations can be used to compare the DL + LL moment to the DL + LL + W moment. The maximum unfactored negative moment due to dead load and live load at the column would be approximately 67.8 kNm (50 kip ft).

Contemporary design

If the project had been designed today and the NEHRP regulations had been used, the north face would have had substantially more inherent resistance to progressive collapse. Had the north face been detailed for a Zone 2A seismic event, the bottom steel at all levels would have required full tensile development lap splices and the columns would have more shear reinforcement.

At a minimum, if the 'notional load' approach of the British codes (BSI, 1972), or if a 1% of gravity minimum lateral load had been a requirement (FEMA, 1995), calculations suggest the progressive collapse would have been mitigated.

Possible failure mechanisms

Introduction

This section summarizes the study of the structural failure mechanism for the nine-storey monolithic reinforced concrete portion of the Murrah Building. The study was carried out to quantify the reasons for the observed failures and to develop options for feasible changes in detail that would reduce the probability of damage from blast and seismic loading in other federal buildings.

Structural data used in the evaluation were obtained from the structural drawings provided by the engineer of record for the building. Nominal material strengths are based on those documented in general notes on the drawings. Design strengths are specified to be 28 MPa (4000 psi) for the concrete and 414 MPa (60,000 psi) for the reinforcement. Tension tests of two samples of #8 reinforcing bars taken from the Murrah Building indicated yield stresses of approximately 496 MPa and 531 MPa (72,000 and 77,000 psi). Tests of three concrete cores from the building indicated compressive strengths of 29, 45, and 38 MPa (4180, 6550, and 5540 psi). These correspond to design compressive strengths of 34, 53, and 45 MPa (4920, 7710, and 6520 psi) when adjusted in accordance with ACI 318 (Corley et al., 1996).

Structural dimensions used in calculations

Figure 14.1 shows the column line designations. Typical column cross-sectional dimensions, and spans for the nine-storey portion of the Murrah Building. Information about the approximate size and location of the crater caused by the blast is provided in (Mlakar et al., 1998).

Arrangement of reinforcing bars in the 914 mm (3 ft) wide by 1524 mm (5 ft) deep transfer girder on column line G at the third floor is summarized in Fig. 14.11. The girder is used to pick up alternate columns and change the column spacing from 6096 mm (20 ft) in the upper stories to 12,192 mm (40 ft) in the lower stories. Reinforcing bar details for the spandrel girders on column line G at the fourth to the ninth floors are shown in Fig. 14.7 and for the roof girder on column line G in Fig. 14.9. Cross-sectional distributions of the girder reinforcing bars are shown in Fig. 14.10.

Calculated nominal section strengths

Tables 14.8 to 14.14 list the calculated flexural strengths of the girder sections, shown in

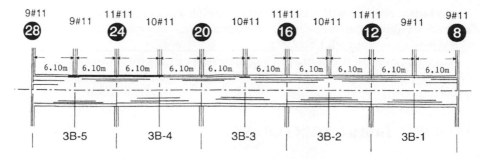

Figure 14.11 *Reinforcing bars in the transfer girder on column line G at the third floor.*

Fig. 14.8, and the data used in the calculations. The flexural strength, M_u, was based on the expression:

$$M_u = A_s f_{sy} d \left(1 - 0.5 \frac{\rho f_{sy}}{0.85 f_c'} \right)$$

where: M_u = flexural moment strength of a single reinforced section;
A_s = total cross-sectional area of tensile reinforcement;
f_{sy} = yield strength of the tensile reinforcement;
d = distance from top fibre in compression to centroid of tensile reinforcement;
r = tensile reinforcement ratio (A_s/bd, where b = width of girder); and
f_c = compressive strength of the concrete.

Table 14.8 *Calculated flexural strength of girder at third floor (steel design yield stress).*

Measurement	Section 16	Section 18	Section 18	Section 20
Width (mm)	920	920	920	920
Effective depth (mm)	1450	1450	1450	1450
Number of bars	11	11	11	11
Bar area (mm²)	1000	1000	1000	1000
Number of bars	–	–	–	–
Bar area (mm²)	–	–	–	–
Sum area (mm²)	11100	10050	3025	11100
Reinforcement ratio	0.0084	0.0076	0.0023	0.0084
Flexural moment strength (kNm)	6140 negative	5625 positive	1775 negative	6140 negative

Note: concrete design strength = 2.75×10^7 Pa   steel design strength = 4.15×10^8 Pa

Table 14.9 *Calculated flexural strength of girder from fourth to ninth floors (steel design yield stress).*

Measurement	Section 16	Section 17	Section 17	Section 20
Width (mm)	460	460	460	460
Effective depth (mm)	815	815	815	815
Number of bars	2	2	2	2
Bar area (mm²)	510	510	510	510
Number of bars	2	2	–	2
Bar area (mm²)	650	400	–	650
Sum area (mm²)	2300	1800	1025	2300
Reinforcement ratio	0.0062	0.0048	0.0027	0.0062
Flexural moment strength (kNm)	725 negative	575 positive	325 negative	540 negative

Table 14.10 *Calculated flexural strengths of girder at roof (steel design yield stress).*

Measurement	Section 16	Section 17	Section 17	Section 20
Width (mm)	460	460	460	460
Effective depth (mm)	1100	1100	1100	1100
Number of bars	2	2	–	2
Bar area (mm²)	400	400	400	400
Number of bars	–	–	–	–
Bar area (mm²)	–	–	–	–
Sum area (mm²)	775	775	775	775
Reinforcement ratio	0.0015	0.0015	0.0000	0.0015
Flexural moment strength (kNm)	350 negative	350 positive	350 negative	350 negative

Interaction relationships for axial load and bending moment capacity calculated, using procedures of ACI 318-71 for column G20, are shown in Fig. 14.12. Tables 14.8 to 14.10 show calculated data made on the basis of design strengths for the materials without strength reduction factors. For girders with #8 and #7 reinforcing bars, a second set of calculations (Tables 14.11 and 14.12) was made based on a yield stress of 483 MPa (70,000 psi) on the premise that the measured yield stresses in bars of that size were credible and provided reasonable evidence to support a yield stress of 483 MPa (70,000 psi). Concrete strength and the yield stress of #11 reinforing bars were assumed to be equal to the design values. Use of the higher values determined from testing would not substantially alter the results.

Table 14.11 *Calculated flexural strengths of girder from fourth to ninth floors (steel measured yield stress).*

Measurement	Section 16	Section 17	Section 17	Section 20
Width (mm)	460	460	460	460
Effective depth (mm)	815	815	815	815
Number of bars	2	2	2	2
Bar area (mm²)	510	510	510	510
Number of bars	2	2	–	2
Bar area (mm²)	650	400	–	650
Sum area (mm²)	2300	1800	1025	2300
Reinforcement ratio	0.0062	0.0048	0.0027	0.0062
Flexural moment strength (kNm)	725 negative	575 positive	325 negative	725 negative

Table 14.12 *Calculated flexural strengths of girder at roof (steel measured yield stress).*

Measurement	Section 16	Section 17	Section 17	Section 20
Width (mm)	460	460	460	460
Effective depth (mm)	1100	1100	1100	1100
Number of bars	2	2	2	2
Bar area (mm²)	400	400	400	400
Number of bars	–	–	–	–
Bar area (mm²)	–	–	–	–
Sum area (mm²)	775	775	775	775
Reinforcement ratio	0.0015	0.0015	0.0015	0.0015
Flexural moment strength (kNm)	410 positive	410 positive	410 positive	410 positive

Limiting strengths for column line G

Three collapse mechanisms for gravity loading of column line G were considered. Calculations were made assuming colunm line G to be two-dimensional with a tributary width of 6096 mm (20 ft) between columns. Effects of increased strength caused by in-plane forces were not considered. The three collapse mechanisms are shown in Fig. 14.13.

Mechanism 1 represents a collapse mechanism for the interior spans of column line G in the as-built condition. Moment capacity at column line 22 is assumed to be finite at the third through to the ninth floors because the bottom reinforcement is continuous, according to information from the drawings. Moment capacities of the girders in the

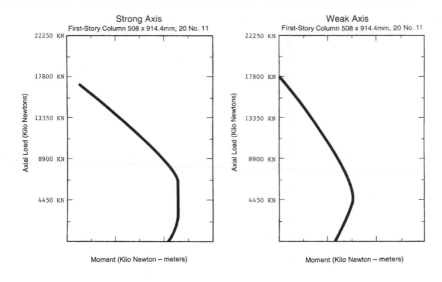

Figure 14.12 *Axial load and bending moment capacity for column G20.*

fourth through to the ninth floors and the roof are assumed to be zero at column line 22 because the bottom reinforcement is terminated at support centre.

Mechanism 2 is admissible if the first-storey section of column G20 is removed. Positive-moment capacity at column line 20 is assumed to be zero at all floors because bottom reinforcement is discontinuous at column line 20.

Mechanism 3 is a check of the possibility of a reduction in strength, with respect to Mechanism 2, related to discontinuities in the top reinforcing bars. As was assumed for Mechanism 2, the first-storey section of column G20 is assumed to have been removed. Negative-moment capacities are calculated at column lines 18 and 22. Positive-moment capacities at column line 20 are set at zero because of the total discontinuity of the reinforcement at that line.

Tables 14.13 and 14.14 show the calculated strengths for the three mechanisms. Option 1 (Table 14.15) refers to all strengths set at nominal design values. Option 2 refers to yield stresses of the #7 and #8 reinforcing bars set at 483 MPa (70,000 psi) and yield stresses of the #11 reinforcing bars and concrete strength assumed to be equal to the nominal design strengths.

Unit load calculated for Mechanism 1, which refers to the intact structure, confirms that the as-built structure had adequate flexural strength for code-prescribed gravity loads.

Comparison of the results for Mechanisms 2 and 3 indicates that Mechanism 3 is not likely to govern. The nominal unit strength calculated for Mechanism 2 is 3 kPa (60 psi), less than 15% of the as-built (flexural) strength of the structure. Considering that the unit weight of the building is likely to be in the range 7 kPa to 10 kPa (150 to 200 lb/ft²), removal of column G20 is concluded to be sufficient cause for failure. The strength

– 249 –

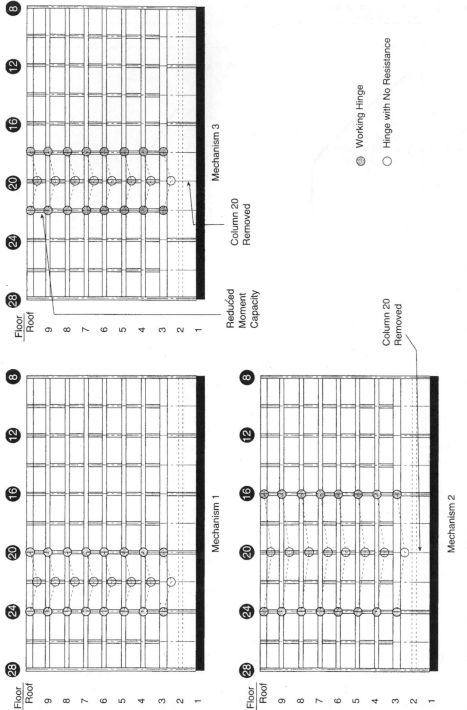

Figure 14.13 *Three collapse mechanisms for gravity loading of column line G.*

Table 14.13 *Calculated limit capacities based on reinforcement yield stress assumed in design (60 ksi).*

Mechanism identification	Yield capacity (Pa)	Yield capacity reduced, $\varphi = 0.9$ (Pa)
Mechanism 1 (as-built structure)	23,500	21,100
Mechanism 2 (column G20 removed in first storey	2900	2400
Mechanism 3 (partial mechanism with column G20 removed in first storey)	4800	4300

Table 14.14 *Calculated limit capacities based on measured reinforcement yield stress (70 ksi).*

Mechanism identification	Yield capacity (Pa)	Yield capacity reduced, $\varphi = 0.9$ (Pa)
Mechanism 1 (as-built structure)	25,400	23,000
Mechanism 2 (column G20 removed in first storey	3350	2900
Mechanism 3 (partial mechanism with column G20 removed in first storey)	5300	4800

Table 14.15 *Calculated strength (unit strength, Pa × 10⁵) of three possible failure mechanisms.*

Mechanism	Option 1	Option 2
1	25,400	23,000
2	3350	2900
3	5300	4800

of 3 kPa (70 psi) calculated for this mechanism with the higher measured yield stress of 483 MPa (70,000 psi, for #7 and #8 reinforcing bars) may be increased another 40% by increases in strength, amount of reinforcement, and effects of three-dimensional response. However, to increase the calculated strength by over a factor of 2 to a level comparable to the self-weight of the structure is not plausible.

The results for Mechanisms 1 and 2 are both hypothetical. For Mechanism 1, strength associated with shear would be exceeded. The calculated load represents a measure of the strength of column line G if its strength were limited by flexure of the girders.

Mechanism 2 assumes that the deformations will not produce secondary moments that cause collapse after failure of the positive-moment hinges at column line 20. The decisive conclusion from the comparison between the result of Mechanism 2 and the dead load is that column line G cannot sustain its tributary weight if any one of the interior columns is removed.

Analysis of results

From the calculations summarized above, it can be concluded that even a 'static' removal of column G20 at the first floor would create structural collapse of column line G between column lines 16 and 24. The events were neither static nor describable by a single variable in one dimension. The failure may be explained as a result of complex interaction of many events in many directions, but Mechanism 2 provides a fundamental and simple explanation for the failure. The structure is not stable without column G20.

It is unlikely that the first-storey column G20, located within 4900 mm (16 ft) of the explosive device (Mlakar *et al.*, 1998) failed in flexure. By the time the deformation required for a flexural failure to occur, the entire cross-section of the column would have been engulfed and the effective pressure on the column would have been reduced. The blast would have 'scoured' the column shell concrete instantaneously. The core, with its heavy longitudinal and light transverse reinforcement, was brittle. Concrete in the core shattered, most likely as a direct result of the pressure-wave impact (brisance). But, even if the initial pressure on one side of the column had been applied slowly with the shell intact, the concrete would have shattered because of the high principal tensile stresses that would have developed almost immediately. There was no possibility of the column absorbing the energy of the impact.

Calculations (ibid) show that columns G16 and G24 at the first floor were also vulnerable to the same failure sequence. However, they could have been destroyed by being pulled down by the falling spandrels. The remaining stub of the transfer girder framing into column G12 suggests that the discontinuity of the top reinforcement would have isolated the failure had the columns not failed first.

For reasons of convenience in placing reinforcing bars, the positive- moment reinforcing appears to have been extended through supports for the spandrel girders at the fourth through to the ninth floors. That detail helped increase the load calculated for Mechanism 1. Had this detail also been used across column lines 12, 16, 20, and 24 for all girders, the load calculated for Mechanism 2 would have increased to 6 kPa (120 lb/ ft^2), based on nominal material strength, which is within range of the self-weight of the ordinary moment frame.

Spiral (helical) reinforcement prescribed for special moment frames by Chapter 21 of the ACI's *Building Code Requirements for Structural Concrete* (1995), but not required for ordinary moment frames, would have resisted the shattering of the column and would have maintained axial-load strength after scouring of the shell. Use of spiral reinforcement, especially in cases with discontinued column arrangements as in the Murrah Building and where detonation of explosives in close proximity is possible, may be an important ingredient for blast resistance.

Blast load analysis

The explosive device was contained in an enclosed truck parked on the paved street along the north side of the Murrah Building (N.W. Fifth Street). Effects of the blast are quantified using methods employed for the analysis of conventional weapons effects on structures (US Department of the Army, 1986).

Calculation of blast loading begins with the estimation of the yield or quantity of explosives detonated. For bursts near the ground surface, the yield or quantity is usually inferred from the dimensions of the crater formed (Fig. 14.14). The engineering survey of the crater forms the basis of this inference. Measurements show the crater was approximately 8.5 m (28 ft) in diameter and 2.1 m (6.8 ft) in depth (Fig. 14.15). Its centre is about 2.1 m (7 ft) east and 4.3 m (14 ft) north of column G20. According to the design drawings and observations on site, thickness of the pavement was 460 mm (18 in) and the underlying soil was dry sandy clay. Information about the truck reported to have contained the explosive device shows the centre of the explosion to have been 1.4 m (4.5 ft) above the ground (Fig. 14.15).

As listed in Table 14.16, the detonation of a spherical charge of trinitrotoluene (TNT) weighing approximately 1800 kg (4000 lbs) at 1.4 m (4.5 ft) above 460 mm (18 in) thick pavement on soil results in a crater whose dimensions are consistent with those measured at the Murrah Building site. In Table 14.16, the crater dimensions for pavement on soil are an average of those for massive concrete and for the dry sandy clay alone, weighted in proportion to the depths of the two materials in the crater. This weighting is substantiated by the results of ongoing research concerning craters in pavements based on soil.

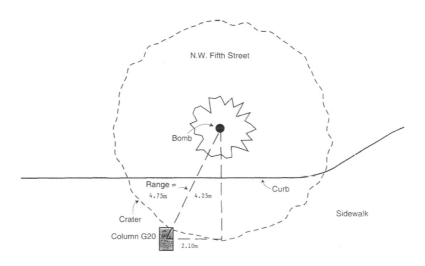

Figure 14.14 *Dimensions of crater.*

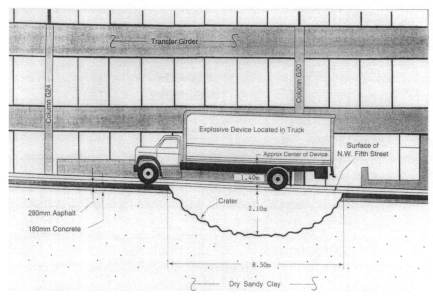

Figure 14.15 *Measurement of crater.*

A detonation also produces an airblast wave that propagates radially from the burst point. This wave is characterized by an instantaneous rise to a peak value termed the incident pressure from which it decays to ambient conditions. Figure 14.16 shows the contours of this incident pressure corresponding to the surface detonation of 1800 kg (4000 pounds) of TNT at the location of the bomb crater. The contours correlate approximately with the level of damage shown for buildings in the neighbourhood. However, this damage is also a function of load modification by nearby buildings and the resistance of the buildings themselves. In particular, buildings shown in the upper part of Fig. 14.16 were not shielded by the Murrah Building. It can be seen that damage in the upper part of the figure is heavier than that in the lower part.

Of particular interest is loading on the nine-storey portion of the Murrah Building. When a blast wave impinges on a structure, a higher pressure, termed the reflected pressure, is developed. The calculated peak overpressures (Department of the Army, 1986) on the north elevation are shown in Fig. 14.17. These range from a maximum of

Table 14.16 *Estimate of yield from crater dimensions.*

Condition	Depth (m)	Diameter (m)
1800 Pa of TNT on massive concrete	0.75	4.0
1800 Pa of TNT on dry sandy clay	2.50	9.5
1800 Pa of TNT – 450 mm of pavement on soil	2.25	8.25
Measured at Murrah Building	2.0	8.5

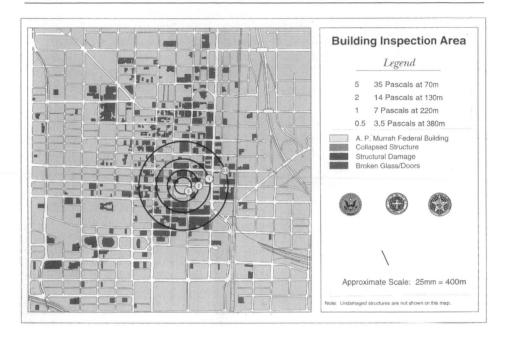

Figure 14.16 *Contours of incident pressure.*

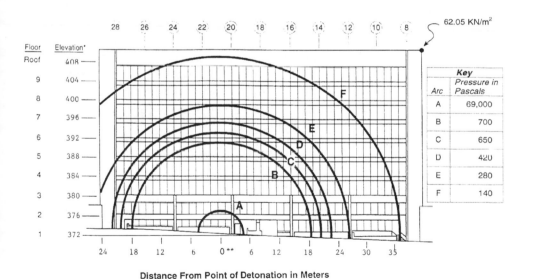

Distance From Point of Detonation in Meters

*Elevation in meters above mean sea level   ** Point of Detonation

Figure 14.17 *Calculated peak overpressures on the north elevation.*

over 69 MPa (10,000 psi) at the point closest to the detonation to a minimum of 62 kPa (9 psi) at the upper west corner, with an equivalent uniform value of approximately 965 kPa (140 psi). While these pressures are extremely large, they acted for a limited duration (Fig. 14.18). The duration ranges from a maximum in the upper west corner to a minimum at the point closest to the blast and has an equivalent uniform value for a triangular pulse of about 5 ms.

An explosive detonation near the ground surface causes a ground shock motion in addition to the airblast loading (US Department of the Army, 1986). For the surface detonation of 1800 kg (4000 pounds) of TNT, the free-field motion at the centre of the building rises rapidly after arrival to a peak value of 230 mm/sec (9 in/s), followed by a gradual decay over the next 270 ms (Fig. 14.19). This ground shock motion is of little consequence to the structure in comparison to the extreme airblast loading.

Blast analysis of adjacent column G20

The bomb crater places the explosive device in close proximity to column G20. The response of structural elements in such proximity is a function of the range divided by the cube root of the explosive mass or scaled range. For column G20, the scaled range

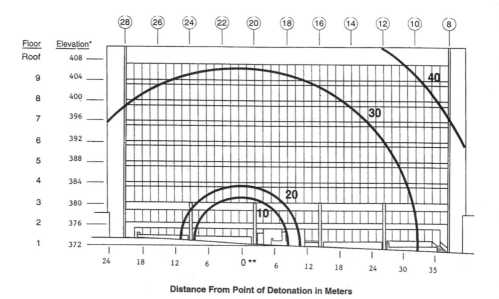

Figure 14.18 *Limited duration of peak overpressures.*

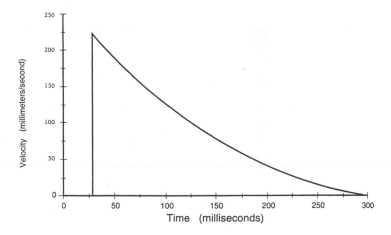

Figure 14.19 *Free-field motion at the centre of the building.*

is only

$$R/W^{1/3} = 4755 \text{ mm} /(1514 \text{ kg})^{1/3} = 677 \text{ mm/kg}^{1/3}$$

or

$$R/W^{1/3} = 15.6 \text{ ft} / (4000 \text{ lb})^{1/3} = 1.0 \text{ ft/lb}^{1/3}$$

in which R is the horizontal range in feet and W is the equivalent charge mass in pounds of TNT.

Based on bomb damage reports from the Second World War (National Defense Research Commiuee, 1946), the destruction of first-storey reinforced concrete columns by the brisant effects of blowing out, severing, and undermining occurs at scaled ranges of 2032 mm/kg$^{1/3}$ (3 ft/lb$^{1/3}$) from cased charges such as bombs or artillery shells. From contemporary research on the difference between the breaching of reinforced concrete walls by uncased and cased charges, the scaled range for destruction by uncased charges is estimated to be 1016 mm/kg$^{1/3}$ (1.5 ft/lb$^{1/3}$). Thus in all likelihood, column G20 was abruptly removed by brisance. This conclusion is supported by the fact that no one whom the team interviewed found any evidence of this column in the debris or in the crater caused by the explosion.

Blast analysis of nearby columns G24, G16, and G12

Column G24 was located outside the range of brisance for uncased charges, but was highly loaded by the detonation. The response of column G24 to this load is approximated as a simply supported beam (Fig. 14.20) between the first- and third-floor elevations (Biggs, 1964). The column did extend below the first-floor elevation to its supporting caisson, but it was not loaded by the blast below this level and received some support from the surrounding soil. The column was also laterally connected at the second floor by a transfer strut, 2B-13, but this feature provided little restraint in the east-west direction excited by the blast.

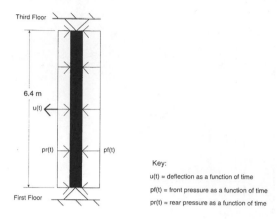

Figure 14.20 *The response of column G24 to detonation.*

The column resisted this loading about its weak axis (Fig. 14.21). Column G24 was reinforced with 20 #11 vertical bars and 2 #4 horizontal ties at 406 mm (16 in) on centre. Strength was limited by the shear resistance at the ends of the column. From the material strengths measured in this study and the estimated axial prestress from dead and actual live loadings, this limiting capacity, V_u, corresponds to 359 kPa (52 psi) uniformly distributed on the 914 mm (36 in) face.

Figure 14.22 shows the blast loading on column G24. On the front face, the load rises abruptly to the reflected pressure, 10 MPa (1400 psi). When the blast clears this face, it falls to the sum of incident and dynamic pressures. Dynamic pressure stems from the particle or wind velocity associated with the blast wave as it strikes structures

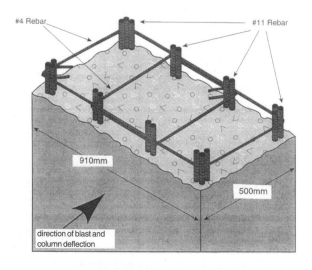

Figure 14.21 *Column G24 resisting loading.*

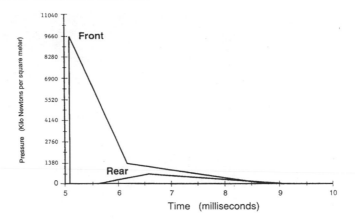

Figure 14.22 *Blast loading on column G24.*

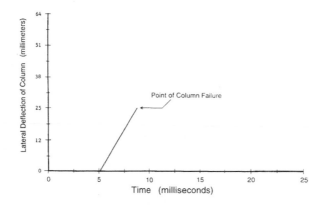

Figure 14.23 *Response of column G24 to blast loading.*

in its path. The blast subsequently arrives at the rear face and rises gradually to the sum of the incident and dynamic pressures at this range and orientation. The effective triangular duration of the net loading is only about 1 ms.

The response of column G24 to this loading (Fig. 14.23) is specifically the lateral deflection of the column at the midpoint, or second-storey level, as a function of time measured after the detonation of the bomb. Notice that most of this response occurs after the net load has diminished to zero so that it is an impulsive structural event. When this deflection reaches 25 mm (1.0 in), the shear at the supports, V, exceeds V_u and the fracture of the element occurs. Because the axial load and corresponding shear capacity are greater at the first floor than at the third floor, shear failure is expected at the top. Immediately after the blast (Fig. 14.24), the upper portion of this column was missing. This physical observation agrees with the results of this analysis.

Figure 14.24 *Upper portion of column G24 missing.*

Figure 14.25 *Incipient brittle failure.*

Table 14.17 *Blast response of intermediate columns supporting north transfer girder.*

Column	G24	G20	G16	G12
Slant range (m)	11.25	6.5	15.25	27.25
Peak pressure (Pa × 10⁵)	95	390	45	8
Duration (ms)	1.3	*	1.7	1.4
Deflection (mm)	55	*	30	5
Shear at supports/ limiting capacity	1.8	*	1.0	0.1

* column destroyed by brisance

As shown in Table 14.17, the slant range (distance from the explosion to the column midheight) is greater for column G16 than for column G24. At a distance of 15 m (50 ft), the peak pressure is still 4 MPa (641 psi). Applying a similar analysis to that performed for column G24, loading on G16 just reaches the shear capacity. This implies an incipient brittle failure which is consistent with the conditions shown in Fig. 14.25.

Column G12 received high direct blast effects from the bomb. It was located at a slant range of 27.1 m (89 ft) as indicated in Table 14.17. Here, the loading was 793 kPa (115 psi). The associated response is only 0.1 of the capacity to resist. Results of this analysis are consistent with the intact condition of this column after the bombing, as shown in Fig. 14.26.

Figure 14.26 *Column G12 intact after the bombing.*

Blast analysis of slabs

Floor slabs in close proximity to the bomb were directly loaded by the blast. The facade of the north elevation consisted of 1520 mm × 3050 mm (5 ft × 10 ft) glass panels restrained by aluminum channels. This glazing offered insignificant resistance to the propagating blast wave. Upon the failure of the glazing, the blast filled the structural bays above and below each floor slab. Filling pressures below the slab were greater than the filling pressures above and caused an upward load on each slab.

This net upward loading is shown (Fig. 14.27) as a spatially uniform pressure. The slab is modelled as a simply supported element spanning from east to west between floor or roof beams (Biggs, 1964). The length of this span between the supporting beams is 4.4 m (16 ft).

The floor slab at mid-span is 152 mm (6 in) deep and reinforced with #4 bars at 229 mm (9 in) on centre in the east-west direction and 457 mm (18 in) on centre in the north-south direction. Because of their location near the bottom of the slab, these bars provide little resistance to an upward loading. Calculated capacity is a uniform load of only 3 kPa (0.40 psi). Roof slabs are typically 152mm (6 in) thick as well but are reinforced with #4 bars at 406 mm (16 in) on centre in the east-west direction and therefore provide a resistance of 2 kPa (0.24 psi) at mid-span.

Figure 14.28 illustrates the loading on the fifth-floor slab between column lines 20 and 22. Loadings are assumed to be the incident overpressures at the range of the midpoint of the structural bays above and below each slab. These loadings are further represented by triangular pulses.

In this particular case, the load from below has a peak of 1 MPa (154 psi) while the load from above is only 8 kPa (87 psi). In both cases, these loads act for relatively short durations.

The response of this slab to the loading is shown in Fig. 14.29. In this case, the static dead and actual live loadings are included as well as the blast loading from Fig. 14.28.

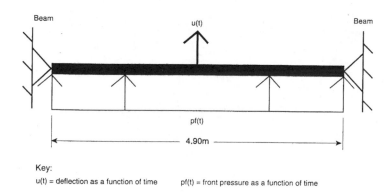

Key:

u(t) = deflection as a function of time   pf(t) = front pressure as a function of time

Figure 14.27 *Model of slab showing net upward loading.*

– 262 –

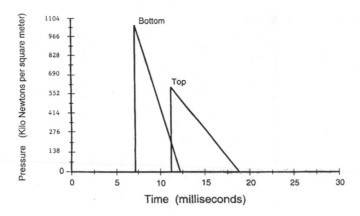

Figure 14.28 *Loading on the fifth-floor slab between column lines 20 and 22.*

Notice that the upward response of the slab has a long period, and the blast event represents an impulsive loading condition. In this case, the maximum deflection is 236 mm (9.3 in). This deflection exceeds capacity of the floor slab and also represents a rotation of 5.3 degrees over the 4.9 m (16 ft) span. Under these conditions, the collapse of the slab from the direct blast loading is expected.

A similar analysis was performed for the other floor and roof slabs in the building (Fig. 14.30). In particular, the slabs in the fifth floor and below between column lines 18 and 24 were sufficiently loaded by the blast to fail as shown. However, the other slabs responded elastically to the differential blast loading and in some cases were not loaded above the static downward capacity.

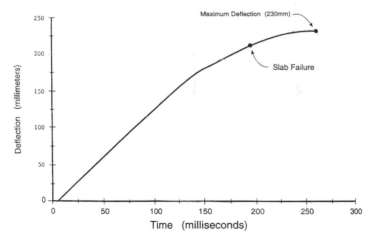

Figure 14.29 *Response of fifth-floor slab to loading.*

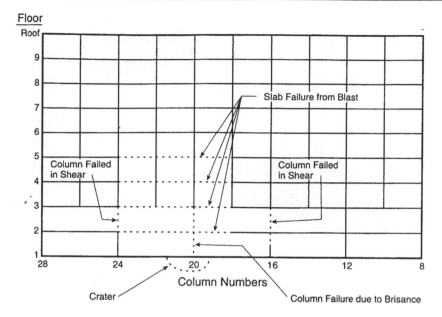

Figure 14.30 *Analysis of other floor and roof slabs in the building.*

Figure 14.31 approximates the inward extent of this directly induced slab failure. Results show a penetration of failure to 12 m (40 ft) at the second floor. Calculations indicate no slab failure at the sixth floor.

Nine-storey portion of Murrah Building

Major structural damage and building collapse occurred at the north side of the Murrali Building, which faced the blast as shown in Fig. 14.32, Here, most of the north half of the rectangular footprint, between columns G12 and G28 (except for the extreme west end), extending 10.7 m (35 ft) into the building, collapsed. Three columns (G16, G20, and G24) supporting the third-level transfer girder were destroyed. The destruction of these columns triggered the progressive collapse of floors above. In addition, between column lines 20 and 24 (a length of 12.2 m; 40 ft), the collapse extended the full width of the building (21.3 m; 70 ft) to, but not through, the south wall. Roughly half of the occupiable space in the nine-storey portion of the building collapsed.

While the north face of the building sustained the brunt of the effects of the blast, structural damage to the remaining exposures was limited.

Ray Blakeney, Director of Operations for the Oklahoma Medical Examiner's Office, has estimated that up to 90% of the 168 fatalities were the result of crushing caused by falling debris.

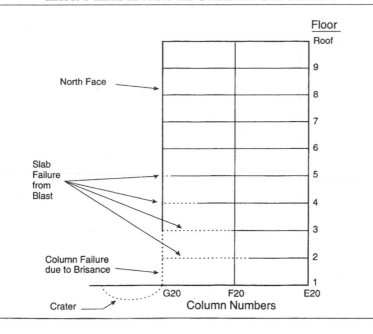

Figure 14.31 *Approximate inward extent of directly induced slab failure.*

Figure 14.32 *Major structural damage and collapse at the north side of the building.*

Analysis and findings

Relevant factors

Based on review of plans, shop drawings, specifications, and construction records, it is concluded that the Murrah Building was designed as an ordinary reinforced-concrete-frame structure in accordance with ACI 318-71 (1971). Records indicate that the building was extremely well detailed.

The structural design was found to have included all of the factors required by the governing building code at the time of construction and to have been very well executed. When this building was designed, there was no requirement to consider earthquake, blast, or other extreme loadings in Oklahoma City.

According to the observations made of the crater and other damage, the blast that damaged the building had a yield equivalent to approximately 4000 lbs of TNT. This extraordinarily large explosion was centered approximately 4.75 m (15.6 ft) from column G20. The blast caused a crater approximately 8.5 m (28 ft) in diameter.

The effect of the blast on column G20 abruptly removed it by brisance (a shattering effect). Loss of this column removed support for the transfer girder on the third floor between columns G16 and G24. Analysis of mechanisms that could result when column G20 was removed shows that an ordinary moment frame would be unable to support the structure above the third floor.

In addition to destroying column G20, force from the blast would cause columns G16 and G24 to be loaded in such a way that yield or near-yield moments would be produced over their lengths from their bases to the third floor. Corresponding shear stresses would exceed the calculated shear capacity of each column. Consequently, calculations indicate that columns G16 and G24 failed in shear. Loss of these two columns would leave the transfer girder unsupported from the east wall of the building to column line G12. Calculations indicate that an ordinary moment frame could not support itself with three columns in column line G missing.

It is noted that the loss of three columns and portions of some floors by direct effects of the blast accounted for only a small portion of the damage. Most of the damage was caused by progressive collapse following loss of the columns.

Possible mechanisms for reducing loss

Ordinary moment frames have limited reserves for dissipating energy from extreme loading such as earthquake and blast. However, special moment frames and dual systems with special moment frames, as defined in the 1994 edition of NEHRP (National Earth Quake Hazards Reduction Program) *Recommended Provisions for Seismic Regulations for New Buildings* (1995) provide structural systems with much higher ability to dissipate energy. It is noted that the NEHRP recommendations for design of special moment frames and dual systems were first available in 1985, approximately 10 years after the building was constructed.

If the more recently developed detailing for special moment frames had been present at the time of the blast, columns G16 and G24 would have had enough shear resistance to develop a mechanism without failure. Consequently, it is likely that G16 and G24 would not have failed abruptly due to the blast loading if special moment frame detailing had been used.

Due to its close proximity to the very large explosive device, column G20 would likely have been destroyed by brisance even if it were detailed as a special moment frame. However, the heavy confinement reinforcement that would have been present would have increased the chances of survival for column G20.

If special moment frame detailing had been used, the following results could have been expected:

1. If column G20 survived the blast, loss of structure would have been limited to those floor slabs destroyed by air blast. This would reduce the loss of floors by as much as 85%.

2. If column G20 was removed by the blast, mechanism 2 (hinging mechanism between G16 and G24) as described in Sozen et al., (1998), would develop. Normal detailing for special moment frame design would provide reinforcement in the transfer girder at the third floor that would greatly increase the possibility that the slabs above would not collapse. Consequently, destruction could be limited to only those areas described in Sozen et al., (1998) as being removed by air blast. Although use of a special moment frame would not completely eliminate loss of portions of the building, it is estimated that losses would be reduced by as much as 80%.

3. If column G20 was removed by the blast and mechanism 2 developed but was not capable of supporting the spans between columns G16 and G24, loss of the structure would be limited to those panels destroyed by air blast and those panels located between column lines F to G and column lines 16 to 24. Resulting loss of floor space to either air blast or collapse would be reduced by more than 50 %.

References

American Concrete Institute. 1971. *Building Code Requirements for Reinforced Concrete* (ACT 318-71), ACT Committee 318, Detroit, Michigan.

American Concrete Institute. 1995. *Building Code Requirements for Structural Concrete* (ACT 318-91), ACT Committee 318, Detroit, Michigan.

Biggs, J. M. 1964. *Introduction to Structural Dynamics*. McGraw-Hill Book Company, New York, NY.

British Standards Institution. 1972. *The Structural Use of Concrete*, CP11O: Part 1 – Design, Materials and Workmanship, BSI, London, England.

Building Performance Assessment Team. (BPAT 1996). *The Oklahoma City Bombing: Improving Building Performance through Multi-Hazard Mitigation*. Federal Emergency Management Agency Report 277, Washington, DC.

Corley, W.G., Sozen, M.A., Thornton, C.H. and Mlakar, P.F. 1996. The Oklahoma City Bombing: Improving Building Performance Through Multi-Hazard Mitigation, *FEMA Bulletin*, **277**.

Corley, W.G., Mlakar, P.F., Sozen, M.A. and Thornton, C.H. 1998. The Oklahoma City Bombing: Summary and Recommendations for Multi-Hazard Mitigation. *Journal of Per-*

formance of Constructed Facilities, ASCE, Reston, VA.

Federal Emergency Management Agency, 1995. *NEHRP (National Earthquake Hazards Reduction Program) Recommended Provisions for Seismic Regulations for New Buildings*, 1994 Edition, Part 1: Provisions. FEMA-222A, prepared by the Building Seismic Safety Council, US Government Printing Office, Washington, DC, 335 pp.

McVay, Mark K. 1988. *Spall Damage of Concrete Structures*. U.S. Army Engineer Waterways Experiment Station, Vicksburg, Mississippi, Technical Report WES-TR-SL-88-22.

Mlakar, P.F., Corley, W.G., Sozen, M.A. and Thornton, C.H. 1998. The Oklahoma City Bombing: Analysis of Blast Damage to the Murrah Building. *Journal of Performance of Construction Facilities*, ASCE, Reston, VA.

National Defense Research Committee. 1946. *Effects of Impact and Explosion*, Summary Technical Report of Division 2, Vol. 1, Washington, DC.

Sozen, M.A., Thornton, C,H., Corley, W.G. and Mlakar, P.F. 1998. The Oklahoma City Bombing: Structural Details and Possible Mechanisms for the Murrah Building. *Journal of Performance of Constructed Facilities*, ASCE, Reston, VA.

U.S.Department of the Army. 1986. *Fundamentals of Protective Design for Conventional Weapons Effects*, TM5-855-l.

15 Failures of masts and towers

Brian W. Smith

Introduction

Through the ages man has sought to build tall structures, ranging from lighthouses and cathedrals to obelisks and windmills. It was recognised that such structures would need to be visually predominant; in the case of the lighthouse to serve its purpose, in the case of the cathedral and obelisk as a symbol and of course the windmill needed to be sited in a high and exposed position to take full advantage of the wind. Within the past hundred years, however, the need for tall structures has accelerated with the requirements for the distribution of electricity and the advent of radio, radar, television and, latterly, mobile telephone communication networks. Advances in the use of materials, design concepts, fabrication methods and construction techniques have led to higher and in that sense more obtrusive structures, but with the penalty of requiring very careful engineering from concept to completion. This has led to lighter structures than ever could have been envisaged even one hundred years ago. For example, the Washington obelisk (169 m high) weighs six times more than the 300 m high Eiffel tower, which at 7000 tons itself, weighs twenty times more than a modern 300 m high guyed mast (*see* Fig. 15.1).

Unfortunately, there has been a history of failures of these light structures, which is high compared with other structures of equal economic and social importance. Brief details of the reasons for this and some case studies are provided in this chapter.

Background

During the last 40 years eight 600 m high masts have collapsed in the United States of America, several 300 m masts in Europe, as well as what was the highest mast in the world – the 646 m long-wave mast in Poland, which totally collapsed on 9th August 1991.

However, including these major structural failures, there have been some 227 reported collapses of guyed masts in Europe and North America, and within the last eight

Figure 15.1   *Typical tall guyed mast.*

months a further 92 cases of structural failures of towers and masts in the former USSR have been reported.

The majority of failures have been caused by ice loading, or a combination of ice and wind. The severity of ice loading on a tower can be judged from Fig. 15.2.

Of perhaps greater surprise, however, has been at least 11 reports in the past 15 years of aircraft flying into the stays of tall, guyed masts. Whilst these have not caused collapse of the structures, the guys were severely damaged; the aircraft did not escape unscathed either! It is now general practice in some Scandinavian countries to design masts for failure of one guy, coincident with relatively low wind speeds, which would provide some security against these incidents or where failure of a guy insulator occurs.

A more sinister development has been vandal damage, which can cause serious concern to owners. The remote location of many of these structures makes protection extremely difficult.

General database of failures

It was realised in the late 1960s by engineers involved in tower and mast design that the

Figure 15.2   *Ice on lattice tower in the United Kingdom.*

failure rate in their specialised forms of structures was considerably higher than in other buildings or structures. There was also a lack of common approaches to their design. Accordingly a group of tower and mast specialists formed a Working Group through the auspices of the International Association of Shell and Spatial Structures (IASS) to address the situation. The collapse of the 383 m high Emley Moor mast in 1969 perhaps formed the catalyst for the creation of this group. Since then the group has expanded (it now has some 80 members) and meets every other year. Whilst one of its aims, that of recommending consistent design directives, has been realised, unfortunately the rate of failure has not decreased.

Table 15.1 shows failures of masts in the past 40 years, based on a schedule prepared by one of the Working Group members (Laiho, 1999) categorised by the type of failure and height. In many cases the cause of failure is not clearly identified. For example, ice loading could have been the primary cause but the failure may have been due to guy rupture. Oscillations could have been given as the cause but these may have been due to wind or wind and ice in combination. In the table 'wind' as a cause has been limited to high wind speed causing overload. However, the cause could have been due to inadequate design or inappropriate materials. Despite these uncertainties the table provides a valuable insight into the likely causes and the preponderance of ice loads as a major cause shows how sensitive guyed masts are to this effect.

The distribution of failures over height of structure does not indicate that the higher the structure the greater risk of collapse, although clearly the majority of masts are probably in the height range of 50 m to 200 m where one would therefore expect the greatest number of failures.

Reports of failures in the former USSR territory have been presented by a member of the IASS Working Group (Roitshtein, 1999). These are listed in Table 15.2 using the same headings, where relevant, for types of failure. For these failures the most frequent cause was due to erection/maintenance problems followed by design/materials. Again it is not clear whether the design criteria themselves were adequate to deal with, say high wind, or whether there was a genuine shortfall in the detailed design of the structure.

Table 15.3 shows both the sets of failures presented in terms of failures in each five year period – where available.

It can be seen that on average some 80 structures have failed every ten years since 1960. During the last three years of records (1996–1998) there have been 29 reported failures in Europe and North America alone – exceeding the average significantly. In addition, of course, there are likely to be many other failures which have not been reported.

Table 15.1 *Failures of guyed masts by cause and height.*

| Cause | Height (m) | | | | | | | | | | | Total |
	0–50	51–100	101–150	151–200	201–250	251–300	301–400	401–500	501–600	601+	unknown	
Ice	13	33	19	20	6	7	11	7	1	1	18	136
Ice and wind	6	7		1	1	1	1	1		2	4	24
Wind	1	2	1	1	1			2	3		1	12
Oscillations		2		1	1	2	4				1	11
Guy failure			1	1		2						4
Outside damage		1										1
Lightning/insulators		1					1	1				3
Erection/maintenance		1						3	2	6		12
Design/materials		1	1		1							3
Plane impact								1				1
Vandals	1	1										2
Unknown		3	8	1	4			1			1	18
Total	21	50	32	25	14	12	17	16	6	9	25	227

Table 15.2 *Russian failures.*

Cause	0–50	51–100	101–150	151–200	201–250	251–300	301–400	401–500	501–600	601+	unknown	Total
Ice	1	1		1		1					1	5
Ice and wind	1			1			1				1	4
Wind	2	1		1								4
Oscillations	1	3	1	2		2	2					11
Guy failure		3				3		1				7
Lightning/insulators		2	1		1	4						8
Erection/maintenance	6	6	1	3	3	4	4					27
Design/materials	1	4	1	4	4	4	1				1	20
Plane impact		1			2	1	1					5
Subsidence		1										1
Total	12	22	4	12	10	19	9	1			3	92

Table 15.3 *Failures by year.*

Period	Failures (excl. Russia)	Russian failures	Total reported
1936 – 1957	Unknown	10	-
1958 – 1960	5	2	7
1961 – 1965	7	12	19
1966 – 1970	20	13	33
1971 – 1975	36	14	50
1976 – 1980	19	15	34
1981 – 1985	45	19	64
1986 – 1990	26	6	32
1991 – 1995	38	1	39
1996 – 1998	29	Unknown	-

Case studies

Emley Moor Mast

Location and layout

The 383 m high cylindrical television mast at Emley Moor was located some 13 km south-west of Wakefield in Yorkshire, at a height of 264 m above sea level. It was constructed in 1966 and designed to meet the requirements of the current wind loading and steel design codes for buildings, as no specific standard for towers and masts was available at that time. The design wind speed was 36 m/s and provision was made for 12 mm radial ice on all members.

Figure 15.3 shows an outline diagram of the mast, its configuration and leading dimensions. The mast was fixed at its base, supported by three planes of guys, at six guy levels. It was constructed on a concrete plinth and, up to a height of 274 m (the fourth stay level), consisted of a 2.74 m diameter cylindrical shell of high yield steel, of wall thickness varying from about 14 mm at the lower levels to about 6 mm at the top levels.

From 274 m to the mast top, the construction was of triangular, lattice form. The face width between 274 m to 335 m (the fifth stay level) was 1.98 m, whilst from 335 m to 383 m (the sixth stay level) the face width was 1.3 m. In the lower lattice portion the main legs were fabricated from 140 mm diameter solid round high yield steel whilst the corresponding members of the upper lattice were made of 127 mm diameter solid round mild steel. Angle sections formed the shear bracing in the lattice sections. The lattice sections of the mast were clad with cylindrical fibreglass shrouds. Between the fourth and fifth stay levels the shroud diameter was 3.65 m, whilst from the fifth stay level to the mast head, the shroud was of 2.74 m diameter.

The structure was provided with internal ladders and, up to the fourth stay level, a personnel lift. External observation platforms were also provided. The stays were steel ropes of locked coil construction socketed at both ends. The stay sizes varied in diameter from 41 mm to 63.5 mm.

The collapse occurred at approximately 17.00 hours on 19th March 1969, in conditions of icing with temperatures around freezing, low cloud and light wind. There was no loss of life or personal injury reported as a result of the failure (*see* Fig. 15.4).

The collapse closed down the owner's (the Independent Television Authority) transmission from Emley Moor. After three days a partial resumption of service using a 60 m temporary mast was made. Full resumption of services took place after 28 days. A Committee of Inquiry was appointed to investigate the collapse (ITA, 1971).

Analyses undertaken

Other than its own weight, the mast was subject at the time of collapse to loading from ice and wind. Owing to the presence of fog and the consequent lack of eye witness accounts it was not possible to establish with any accuracy the distribution and intensity of ice loading on the mast and stays. However, such evidence as was available indicates

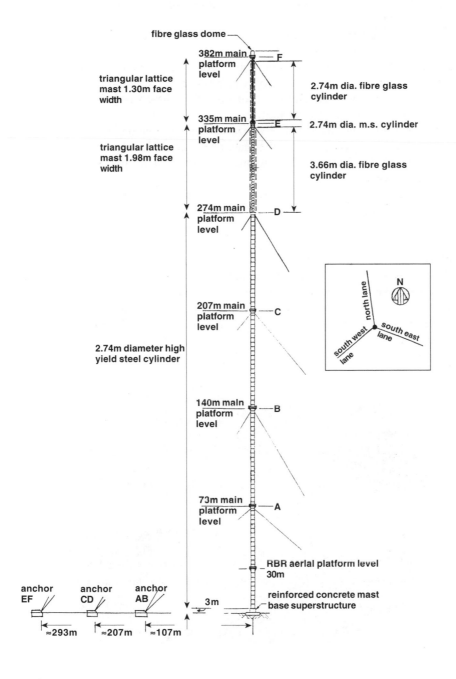

Figure 15.3 *General arrangement of Emley Moor mast.*

that, at the time of collapse, the mast top was probably deflected northwards by the presence of more and denser ice on the stays in the northern and southeastern lanes than on the stays in the south-west lane.

Analyses were undertaken under a variety of assumed distributions of ice which would have produced overstress in the structure sufficient to cause collapse; however it was concluded that none of them alone would have produced both the characteristics of the local failures and the configuration which was observed in the wreckage on the ground. It was concluded therefore that collapse of the mast was not occasioned by ice load alone, although it was a contributory factor.

From such wind data as were available it was deduced that the mean fifteen second wind speed over the upper half of the mast on the day of collapse probably ranged between about 8.5 and 10 m/s. The pressure effects of this wind as a static force were of negligible proportions.

The mast had been subject to aerodynamic excitation by vortex shedding both during construction and on several occasions between its completion and the collapse. During the last four days of its life it was subject to a wind of notably steady direction and velocity, resulting in severe oscillations along an axis lying approximately north-south. Wind speeds were such that, during some of this latter period, if not all, compound modes of flexural vibrations were probably occurring.

The structure was analysed for dynamic response to vortex shedding and it was found that combined modes of oscillation would cause high stresses in the lattice portion of the mast between the 305 m and 320 m levels where, it was believed, the failure initiated.

Figure 15.4 *Collapse of Emley Moor mast.*

Cause of collapse

According to the calculations undertaken, the stresses resulting from dynamic excitation augmented by those arising as a result of asymmetric icing, would exceed the tensile strength of the flange joints of the southern leg in the 1.98 m lattice section of the mast. From examination of the wreckage it was concluded that initial fracture occurred in one of these joints at the 313 m level, most probably accompanied by partial fractures in the joints immediately above and below (319 m and 307 m).

Calculations and laboratory tests showed that prior fatigue damage, which would arise as a result of fluctuating stresses produced by aerodynamic excitation during the service life of the mast, may well have played some part in reducing the tensile strength of the critical flange joints. However, the evidence available from the wreckage neither corroborated nor contradicted this theory.

Consequent upon the initial failure, the top section of the mast became overstressed and, under the effects of its northerly motion in oscillation combined with the release of strain energy, and probable guidance by contact with the lower stays, landed about 150 m due north of the mast base. The falling wreckage of lower portions of the lattice and its associated guys brought about the progressive collapse of the remainder of the mast, which fell, more or less vertically, in the vicinity of its base.

Ylläs Mast, Finland

Location

The mast that collapsed at Ylläs was a 212 m high guyed lattice mast located in the municipality of Kolari, Finland and was erected in 1968. The mast was designed to Swedish standards current at the time, which differed marginally from those for Finland. Ice load was not then specified in the Finnish regulations. The base of the mast was at an elevation of 697 m above sea level. Figure 15.5 shows the general arrangement of the mast. Failure occurred on 23rd November, 1970, following which a Committee of Inquiry was appointed by Oy Yleisradio AB (1972), the mast owners and operators, to investigate the collapse.

Performance of the mast

The materials, fabrication, hot dip galvanising and installation of the mast were inspected on the basis of the structural drawings and the specifications and were found to conform to the required standards.

Strong vibrations in the mast occurred in the autumn of 1968 and a temporary damping system was installed in March 1969, directed at the vibrations which at that time appeared to occur solely in the stays.

It was observed that the mast iced up more than expected during the first two winters of service. Discussions were held with experts in the autumn of 1970 in order to clarify this phenomenon and examine the static properties of the mast. This study was not completed due to the collapse of the structure.

During the first year inspection, carried out in September 1969, no damage such as loosening of the bolted joints was observed. A clear change, however, had occurred in

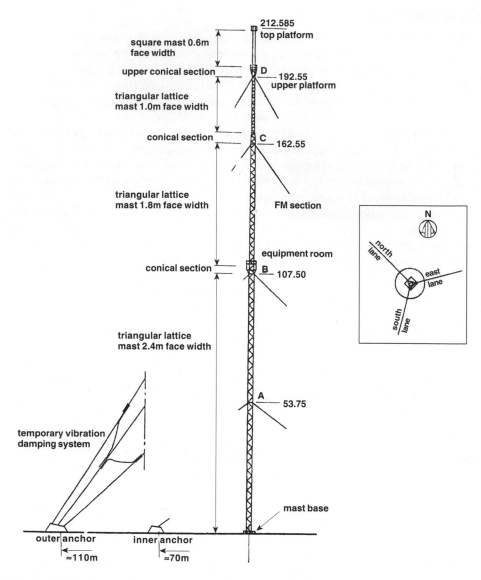

Figure 15.5 *General arrangements of the Yllas mast.*

the initial tension of the stays that had, with the exception of stay level C, decreased by about 1 to 3 tons. This reduction in still air tensions was attributed to the strong vibrations noted above. The initial tensions were corrected and the mast was straightened. On the basis of the inspection, the mast was approved for acceptance but it was insisted that any further vibrations should be observed and an effort made to develop a permanent effective damping system.

Cause of collapse

The strong vibrations observed at the instant of collapse, combined with the heavy load of ice, were the probable reasons for the connection bolts of the southern main member of the upper conical section below stay level D to fail. Recorded variations of the received field-strength in the radio link before the collapse and strong vibrations in the mast observed during previous years also support this theory. On the basis of both static and dynamic analyses, the tensile strength of the joint at the point where the break presumably occurred proved to be the weakest of the mast. This was true for several different calculated loading combinations. The strength of the joint might have been lower than assumed due to loosening of the bolts. (Inspections of other similar masts had shown that nearly one-third of the bolts examined were strained and hence loose.) If three of the eight bolts at the critical section had been previously overstressed then the wind and ice load conditions alone at the moment of collapse might have caused the joint to fail.

On the basis of the location of the parts of the fallen mast, the sequence of failure was predicted. The collapse began when the bolts connecting the southern main member with the upper conical section failed below the uppermost stay, due to the effects of the wind and the eccentric ice load, augmented by the vibrations. The collapse continued from there as a chain reaction, all the way down to the base of the mast. It was possible to discern three separate phases in the collapse:

- In the first phase the impact stress caused by the upper part of the mast falling, sheared the anchor bolts in the base of one of the uppermost stays thus causing the lower end of the stay to come loose and the section of the mast located between the two uppermost stay levels to rupture.
- In the second phase, the top mast section that had broken off slipped down along the eastern stays attached at levels C and B, breaking the parts anchoring the stay and loosening the lower end of stay B. The loosening of the stay exerted a powerful horizontal force on what was left of the mast. This force broke the parts of the mast between the lowermost and the third lowest stay levels.
- In the third phase those parts of the mast that fell to the ground damaged the base of the mast and also caused the lowermost part of the mast to collapse.

Teutoburger Mast

Location and layout

The 295 m high mast (Fig. 15.6) was located on the Teutoburger Wald mountain range near Detmold and was built in 1970. It was constructed of a 230 m tubular steel column topped out by a further 65 m cylinder of glass fibre construction, cable-stayed at five levels of three arrays of guys set 120° apart.

Critically, as it proved, during manufacture slots had been cut into the undersides of each of the three stay plates of the top guys to the mast. The slots were needed to make the plates fit over the lower of two outstand rings stiffening the top and bottom of a local thickening tube bolted to the mast cylindrical column (*see* Fig. 15.7).

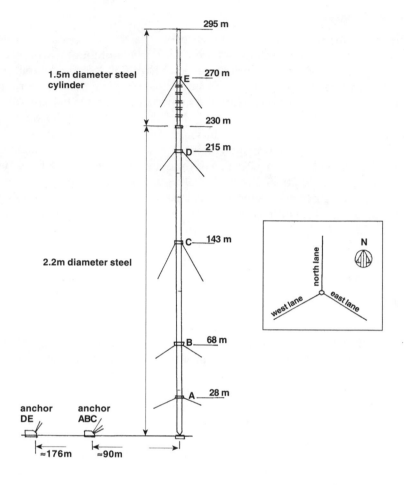

Figure 15.6 *General arrangement of Teutoburger mast.*

Failure occurred on 15th January, 1985. A temporary 85 m mast was quickly installed to maintain services and the structure was finally replaced by a 300 m mast in 1986. A technical investigation into the failure was undertaken for the owners of the mast, Westdeutscher Rundfunk (Krug, 1986).

Performance of the mast

It was considered that the mast suffered significant vibrations for the two years preceding the collapse. The position of the mast on top of the narrow Teutoburger Wald range led to a critical change in the behaviour of the prevailing easterly wind, with the speed being increased on the windward side of the range and lowered on the leeward side, resulting in a 'hydraulic fissure or jump' being created at the summit near to where the mast was positioned.

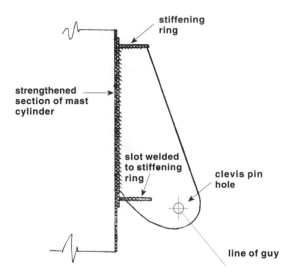

Figure 15.7   *Stay plate on Teutoburger mast.*

The mast stood for a period in the area of the hydraulic fissure/jump immediately preceding the collapse and for a lot longer in the strong turbulence left behind by this unusual atmospheric event.

Cause of the collapse

In the conditions of the atmospheric inversion, a broad frequency range with extremely high energy was present as a stimulus for extremely fierce vibrations of the stays; conditions which lasted about 12 to 15 hours.

The resulting cable vibrations led to the failure of one of the uppermost stay connections at the 270 m level (*see* Fig. 15.7). The fracture of the stay plate occurred because of continually fluctuating strains in different directions causing cracks in the welding seams and material of the junction plate.

The mast had no system redundancy and with failure of the stay plate the structure lost its stability.

Following the failure at the 270 m level, the freed length of the mast then fell southwards, bending about the 'D' stay level at 215 m, losing its uppermost fibreglass section and swinging down to hit the mast at the 161 m level. There the mast broke and the whole top section catapulted northwards, but fell on one of the stays at the 143 m level, tearing its connection plate from the mast and so freeing the length from 161 m to 70 m which crashed to the ground again in a northward direction. This falling section then damaged the north stay in the second level at 68 m and went on to hit and snap the bottom level stay. The loss of tension at 28 m in the northern stay combined with the now unbalanced tension at the 68 m level above, led to the final section of the mast kinking and falling southwards.

Gabin Mast, Poland

Location and layout

This lattice mast was located at Konstantynow, Gabin and, prior to its collapse, was the highest mast in the world at 646 m. It was a long-wave broadcasting structure erected in 1975. The general arrangement of the mast consisted of a triangular lattice of constant face width of 4.5 m throughout its height (Fig. 15.8). The base of the mast was pinned and supported on a ceramic base insulator. The mast collapsed on 8th August 1991. Ciesielski (1992) describe the cause of the collapse and Zoltowski et al (1992) provide a general description.

Performance of the mast

The mast operated perfectly well in service and the transmitted programme under the name 'Warszawa 1' was heard almost all over Europe. In good conditions it reached even the northern parts of Africa and Asia Minor. It is worth mentioning that within its reach there were about 4.5 million ethnic Poles living abroad and many Poles staying temporarily or working abroad.

The mast was supplied with a passenger lift, ladder and platforms, but being a radiator no antennae were mounted on the structure.

Cause of collapse

In 1991, after 15 years service, it was decided that the mast guys should be replaced. The work started from the upper three stay levels and was undertaken in relatively calm conditions. To maintain stability, temporary stays were connected to the structure through auxiliary stay plates. It appeared that during the exchange of the penultimate level of guys, bolts were inadequately tightened. The riggers had descended from the mast and, after some short time, the mast collapsed. Fortunately there was neither loss of life nor injuries.

Other failures

Ice failures

The failure of very high guyed masts (over 305 m high), due to icing, was studied by Sundin and Mulherin (1993). For thirteen of these masts reasonable meteorological data in the vicinity of the sites were available and the failures covered a wide area of the United States over a period from 1973 to 1991. Failures of three masts in Scandinavia were also studied. It was concluded that the failures were found to be the result of both precipitation icing and in-cloud icing. The precipitation icing, with freezing rain as the main type of precipitation, dominates the failures in flat terrain with high winds. Such terrain is typical in the mid-west United States of America, where there were many collapses. This terrain and form of icing events was also found in Sweden and Finland. One difference however is the frequent and large changes of temperature in the mid-west of the United States of America.

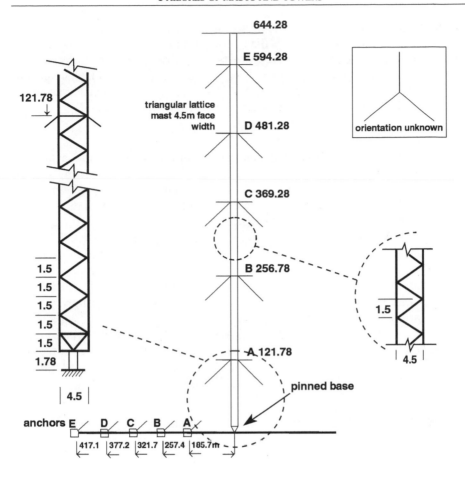

Figure 15.8 *General arrangement of Gabin mast.*

During the days before any failure frequently there was some kind of weather change. In some of the cases it occurred a few days prior to the failure day and in others it occurred on the day of failure.

It is evident that all of the events cited in the report were preceded by warm frontal-type weather conditions. Weather associated with a warm front is characterised by prolonged periods of complete cloud cover, lowered ceiling height and increasing winds, which generally shift from easterly to southerly. Precipitation tends to change from dry snow to wet snow and freezing rain.

These conditions set the stage for collapse by loading the mast with large masses of ice, usually high density glaze from wet precipitation or hard rime from in-cloud icing with high liquid water content. These conditions led to eleven of the sixteen cases where the resulting mast accretions were heavy to severe. However, on some masts the total accretion was only moderate, which leads to a second consideration for mast failure in ice conditions. If severe ice loading is the first step toward failure, a triggering mechanism is often the final step. Triggers are a variety of occurrences that are exacerbated by the ice accretion. Examples of triggers in the selection of failures studied are incomplete construction, brittle steel and fatigue-sensitive details, loose bolts, cable and mast oscillations, dynamic shedding of the ice, and increasing/shifting winds.

Maintenance failures

There are many examples of failures during maintenance operations. These include the collapse of a 600 m guyed mast at Houston, due to the failure of a clamp used in hoisting a 6-ton antenna on to the mast. This collapse killed five men who were on or attached to the antenna and injured three others who were working on the roof of a nearby building. The mast had recently been constructed and the work crew were installing the FM antenna, when the clamp broke allowing the antenna to fall. It sliced through one of the mast stays at about the 300 m level. This was sufficient to fail the mast, which collapsed completely.

Another case was an almost new 610 m high lattice mast in east Missouri, which collapsed when workmen were fitting new bracing rods in the structure to replace the original members which had been found to have minute cracks in the heat-affected welds. Three workmen were killed. The rods were being removed one at a time, using a chain winch to provide temporary bracing. It was believed that cracks in two of the original rods coincided with sudden slippage of the winch at the 146 m level. Prior to the collapse 800 rods had been replaced, starting from the top of the mast.

Conclusions

The high rate of failure of these specialised structures underwrites the need to treat their design, fabrication, construction and maintenance extremely carefully. Unfortunately, particularly in the United States of America, many of these structures are ordered by clients whose knowledge of the structural requirements is relatively limited. Initial-cost economics dictates the choice of structure and contracts are awarded on a design and build basis at minimum tender price. The cost of the structure is usually relatively low compared with that of the antennas it is supporting and frequently receives less attention from the owner. Of concern in the future is that existing structures are being used to support ever-increasing arrays of aerials and antennas. This is due to the fact that increasingly it is becoming difficult to obtain planning permission to erect new towers and masts – for very good environmental reasons. However with the staggering increase in the use of telecommunication facilities – broadcasting channels, microwave networks and mobile telephone usage – greater pressure is being exerted on owners of towers and masts to mount more aerials on their existing structures. Extreme care is necessary to ensure that this does not precipitate even more failures, either during conditions of potential overload from wind or combined wind with ice, or during periods of installation or maintenance of the aerials.

References

Ciesielski, R. 1992. On breakdowns and defects of steel radio-television masts. *Inz. I. Bud.* Nr 3 (in Polish).

Kilpi, J., Laine, J.J., Olkkonen, E. and Vuorenvirta, M. 1972. *Board of Inquiry into Collapse of Ylläs Mast*, for Oy Yleisradio AB.

Krug, S. 1986. *Report on failure of Teutoburger Mast*.

Laiho, J. 1999. *Some known mast failures*. Database produced for IASS Working Group 4:Towers and Masts.

Roitshtein, M.M. 1999. *Analysis of mast and tower failures*. Meeting of IASS Working Group 4: Towers and Masts, Krakow.

Sundin, E. and Mulherin, N. 1993. *Icing-related tower failures in the USA and Fenno-Scandinavia*.

Wex, B.O., Flint, A.R. and Weck, R. 1971. *Investigation into the Collapse of the Guyed Cylindrical Mast at Emley Moor*, for Independent Television Authority.

Zoltowski, W., Klesta, L., Gutkowski, W., Niewiadomski, *et al.*, and Pietrzak, L. Papers on causes of the mast collapse at Gabin. *Inz. I. Bud.* Nr 9, 1992 (in Polish).

16 Precast concrete cladding and structural integrity

J.N.J.A.Vambersky and R. Sagel

This chapter demonstrates what can go wrong when properly authorized and responsible structural engineers are not employed on a construction project to ensure that all structural links fit in place and that structural integrity is secured.

A building in the Netherlands illustrates disturbing developments with respect to the minimum work required in the preparatory phase of the building process. It seems that in particular the structural engineer is being pushed out of this important part of the building process by counterproductive economies of some building process managers and/or developers, who do not recognize the importance of the structural design and structural integrity of the final product. Especially in today's building industry, where different structural elements are supplied by different subcontractors, there is a great temptation also to subcontract to these parties the relevant parts of the structural calculations. The assumption is that by combining the constituent parts, an adequate structural design will result but without the need for a structural engineer or his fee.

This method of working may seem cheaper, as the cost in man-hours to make the required calculations and drawings is apparently removed. However, this does not mean that those costs do not exist and, in the end, will not be paid indirectly by the principal. The one thing that will certainly be lost is independent structural design and checking of the structural integrity of the total project.

What the consequences of such an approach may be will be illustrated with reference to a building where a number of 'imperfections' appeared. The building was designed and constructed without the involvement of an authorized and independent structural engineer. Even the architect was not involved in the process after he had made the final sketch design. The opinion of the developer was that the architect was no longer needed. The different contractors, sub-contractors and suppliers were considered to be capable of completing the remainder of the work which would have been done by the architect. The same applied to the work of the structural engineer. His work and responsibility was distributed amongst a number of different subcontractors and manufacturers of precast concrete elements.

The sub-contractor responsible for the pile-driving was assigned the loading calculations for the building. The contractor for the basement was asked to make the appropriate calculations for various elements but not their overall structural design. Finally, the drawings of the upper structure were made the responsibility of the different manu-facturers of the precast concrete elements. In this way, the structural design was consid-ered to be solved – and solved in an inexpensive way. It was considered that the final sketch design of the architect gave sufficient information for the suppliers of the precast concrete elements to do their job.

Obviously in such a process a specialist manufacturer is employed to supply the precast concrete floors. The floor elements, beams and columns will each be ordered from a supplier who gives the lowest price. The same applies to the load-bearing inner façade and the precast architectural concrete. A marginal reduction in the cost per square metre of façade can make a significant difference. In this way, the cheapest components can be obtained and the costs of preparation and design can be limited. A building with over 10,000 m² gross floor area can in this way be constructed rapidly.

The main structure of the building was erected within a reasonably short period of time, which is usual when a building structure consists of precast concrete elements (Fig. 16.1). The cavity wall insulation was placed and the first architectural precast concrete elements of the façade were imported to be fixed in place. The ground floor elements in front of the columns were erected first to be followed by the horizontal parapet elements of the first floor. The parapet elements have two concrete ridges by which the elements are hung onto the load-bearing inner façade walls. In these walls block-outs are provided for this purpose (Figs. 16.1 and 16.2). At the bottom of the parapet elements are two guard bars for fixing the element and adjusting it into the correct position. It is a well-tested concept that ensures a quick and easy assembly, since

Figure 16.1 *The wing of the building in question viewed from the inside. The block-outs in the bearing inner façade walls that are left out for the ridge supports of the architectural precast concrete cladding, are clearly visible.*

Figure 16.2 *Detail of a ridge support.*

the elements have a flat surface. When the first parapet element was put in place, the ridges bearing the self-weight of the element broke off (Figs. 16.3 and 16.4). The same occurred when the second element was placed. How was this possible? The precast concrete concept was simple and chosen carefully!

The only solution was to call in a consulting engineer to examine the situation. When a parapet element of the façade was inspected, it became clear that the ridges were reinforced well, but the slab did not have any extra reinforcement at the ridges – merely the standard mesh reinforcement. However, at these points, the slab has to bear the same moment as the ridges plus the forces of the support reaction. This vital extra reinforcement in the slab was not included, nor did it appear in the shop drawings. Unfortunately, at that moment, all 10,000 m² of façade elements had already been produced and were awaiting transportation.

Fortunately, an inventive solution proposed by the consulting engineer reduced the damage substantially. The solution was simple and effective. As the building does not exceed four storeys, it was possible to stack the elements of the façade upon each other. The stacked elements at ground floor level were supported by cast in situ concrete corbels which were applied afterwards to the basement walls and foundation beams with the use of drilled anchors. In this way, the ridges remained free from their supports and unloaded. The remaining fixings and other details were flexible enough to permit vertical movements of the façade due to changes of temperature without being damaged. In this way, all elements could still be used.

If all the elements had stayed in place during the assembly but then failed during a storm, the consequences could have been disastrous. The Netherlands has a good reputation in the use of precast concrete elements which was blemished by this event.

Figure 16.3   *Parapet elements of the façade containing failed ridges.*

Figure 16.4   *Detail of a failed ridge.*

When visiting the site in connection with this façade disaster, the consulting engineer came across some other disturbing structural problems which raised more worries and questions in addition to the matter of the façade.

The building, with a gross floor area in excess of 10,000 m², was erected using precast concrete load-bearing façade elements (inner cavity sheets) and pre-stressed precast concrete hollow core slabs which carry from one load-bearing façade to the other. The two-level basement is used as a car park. The basement was cast in situ and the upper storeys completely pre-fabricated. The floor plan of the complex shows a number of wings at 90° to each other. Above the basements, the building is three to four floors high. In principle, the load-bearing façade elements run through to the ground floor (Fig. 16.5). Due to the functional demands such as the minimum dimensions for the entrances and driveways in the car park, it is not always possible to continue the chosen uniform grid of load-bearing façade at all locations at ground floor level. Entrances and other large vertical support-free elements are very often required at this level, as was the case in this building. The entrance of the car park required a bigger opening than the standard grid of the load-bearing façade allowed. Because of this a transfer structure using columns and beams was introduced into the load-bearing façade (Fig. 16.5).

The actual sequence of events can no longer be traced, but seeing the results (Figs. 16.6 and 16.7), the following was most probably the case:

1. In the final sketch design, the load-bearing façade was drawn within the standard thickness, without the necessary fine tuning such as enlarging of the column dimensions at the position of the transfer structures, etc.
2. Taking the final sketch design drawings from the architect as a starting point, the construction of the basement area was worked out by the building contractor

Figure 16.5 *Load-bearing façade with the car park entry.*

responsible for that particular part of the building. Thus the dimensions of the protruding part of the basement wall which was to carry the load-bearing precast concrete façade were based on the thickness of this load-bearing façade. This can be clearly seen from the photos (Figs. 16.6 and 16.7).

3. In addition to the large openings in the façade, the remaining vertical components between these openings could not carry the weight of the building above, but this only became apparent when the manufacturer of the load-bearing façade elements started to make his calculations. He computed the calculations correctly and changed the precast elements at these points into the previously mentioned columns and beams. As to how the constructions underneath these elements was to appear was neither his concern nor his responsibility for it was not part of his commission. Possibly, he thought that any details concerning the support of the precast concrete columns would be adjusted to the final dimensions of these columns.

A check was made on the bearing capacity of the structure. As there were no original calculations to be found, which would have shown that the eccentric and reduced supports of the columns had been taken into consideration, a new calculation was made. From this calculation the vertical load-bearing capacity of the columns as well as the column supports was found, fortunately, to be adequate.

Figure 16.6 *Result of lack of coordination between basement and precast upper structure.*

Figure 16.7 *Detail of the precast column bearing at the car park entry.*

The question remained, however, whether the connection between the precast column and the in situ cast underlying ridge was adequate. Good engineering practice demands that such a connection should be able to guarantee structural integrity even in the case of accidental loadings. In this respect, one doesn't have to look very far; one of the columns stands just in the middle of the exit and entrance of the basement car park (Figs. 16.7 and 16.5). The chance of a car colliding with the column is far from small, which is evidenced by the coloured traces of such collisions in car parks generally.

In this project, such a collision was not taken in account, either in the calculations of the detail, or in the form of any preventative measures taken. The client had been warned of this situation, so that he was able to take the necessary precautions. With a well-placed crash barrier, essentially the problem was solved. However, it remains a fact that without the original problems concerning the façade elements, no attention would have been paid to this deficiency. The suitability of the column to carry its load would then have been dependent only upon the driving skills of the car park users.

Should the column have received an impact and been knocked out, then in all probability most of the load-bearing façade would have stayed more or less in place. However, the precast concrete beams directly supported by the column, and the precast hollow core slabs resting upon them would have lost their support and fallen down. For the tenants of that floor and the driver and his car underneath, the consequences could have been dire.

Due to the inadequacies in this particular project, it was no surprise that many additional problems were discovered. These varied from a seriously leaking basement, as a result of substandard installation and poor detailing of the expansion joints, to the non-existence of shear reinforcement. Insofar that these discrepancies were brought to light, they were corrected afterwards by injection and application of steel plates glued to the concrete surface to cover the function of missing shear reinforcement. All of this remedial work was very expensive and could easily have been avoided.

The old proverb, 'Look before you leap' is quite apt in this respect. 'Thinking twice' at an early stage of the building process will be rewarded as time proceeds and obviously the design phase is important in this respect. After all, it is during this phase when one should stop and think 'how and why' and 'what should be done and what should not be done'. During this initial phase, decisions are made on expenditure and savings. Savings made on design work, or even omitting completely the input of certain specialists during this phase can be counterproductive, ultimately more expensive and possibly dangerous. 'Look before you leap', is certainly applicable to the construction of buildings; not only for economical, but also for safety reasons.

Index